Tipping Point
The Coming Global Weather Crisis

Revised Edition

By Michael J Little

MARCH 29, 2015

Second Edition

ISBN-13: 978-1456455255
ISBN-10: 1456455257

TO

MY CHILDREN

AMY, JOE, ANNA, AND JAMES

It is our moral and ethical responsibility to leave the world in a better condition than we received it.
Or at least not screw it up so badly that it is unlivable for our children's children.
We have not been doing a good job.

CONTENTS

I FORECAST

One day when the Arctic Ocean is free of ice
(an event that should happen in next 30 years),
an ice storm will hit the Eastern Canada and the Northeast;
after the ice falls, it will not melt.

This will be the first year of little to no summer in portions of the northern Hemisphere.

If man does nothing, it will be over 36 million days before summer comes again
(that's approximately 100,000 years)

FORWARD

I do hope that I am wrong, but if I am correct this may be the most important book ever written. But only if it gets widely read and people understand the magnitude of the threat that impending climate change may bring.

Six years ago I started working on Tipping Point and stated the warming Arctic eventually breaks the Jet Stream. At the time no one talking about climate change or a weakened jet stream, both of these are now accepted knowledge and playing havoc with our climate. We are certainly at the tipping point, the first step towards dropping into a new ice age, from here things happen pretty quickly.

Debates continually rage between liberal and right wing fractions of government, scholars, scientists, and the general public about the weather extremes over the last few years. This comes at a time when our ineffective politicians are completely deadlocked. No two groups can agree on anything and they usually resort to name calling between the two groups. I am an independent researcher, financially beholden to no one, and don't have a dog in this fight; but I do see baffling events. Some of these:

1. Typhoons developing very near the equator. As a weather forecaster in the Navy for 20 years I spent most of my time in the west Pacific and tracked literally hundreds of storms. They usually have a set pattern and this pattern has been changing. They are developing further south, almost literally on the equator. This is insane, they shouldn't do that.
2. Areas that have not typically seen tropical storms are now at risk, endangering millions of people around the globe. Increased accuracy of weather forecasts have resulted in far fewer deaths than in years past, but property damage from storms, hurricanes, and storm surges have caused billions of dollars in damage in the US alone.
3. Warm currents are extending much further northward into regions such as the Gulf of Alaska in the Pacific and Newfoundland in the Atlantic. Surface and subsurface fingers of these currents are extending into the Arctic and hastening the melting of the Arctic ice pack. Both the Arctic sea ice and Antarctic glaciers are melting from the top down and the bottom up.

4. In the Arctic while the area of new ice has actually increased in size over the last few years, the warmer summers are melting all of it; resulting in much less ice that survives through the summer. This results in a gradual thinning of the ice that survives year to year, which has been backed up by numerous studies. The area of coverage of multiyear ice is a fifth what it was a generation ago and a quarter of the thickness.
5. The Jet Stream is weakening, resulting in heat waves and droughts in the west and colder and more intense winters in the east (I forecast this 5 years ago).

An erratic Jet Stream is the reason we slide into Ice Ages, it lets cold air masses stagnate over our continents until they are so cold they get locked in place and ice begins to build.

So what is going on? It is really something we have not seen before?

In early 2010 I published the previous edition of Tipping Point and forecast continued warming until 2020, followed by rapid cooling. I based this on the current warming trend and the traditional solar sunspot cycle. What I did not know was the sun is going into a period of less sunspots. This last happened 200 years ago during the Dalton Sunspot minimum, historically this period in the 1800s was tied to global temperatures much less than normal and many years in which summer simply did not come; both the Thames and Hudson rivers froze over during many years in this period. Granted numerous extremely strong volcanic eruptions did not help things at all.

It now appears certain we are going into another period of minimum Sun Spot activity. The last Sun Spot Maximum that peaked in April 2014 was very weak, much less than the previous two maximums. We now have a steady trend of 15 years of decreasing Sun Spot activity. At the same time we have seen global temperatures not rising nearly as much as expected (leading many to decry that Global Warming is a scam and the amount of impact that greenhouse gases is minimal). Only time will tell if the sun is actually going into a rest period, if it does we can expect a hundred years of much colder temperatures than we have enjoyed over the last hundred years.

Today, we have a situation where atmospheric greenhouse gases are increasing to levels higher than the start of the last ice age (128,000 years ago) resulting in more of the incoming heat from the sun becoming trapped, as it cannot radiate back out to space at night as long wave radiation (infrared heat). Our warming trend has been offset as the suns output for the last 15 years has been slightly less than expected, causing temperatures to not rise as rapidly as pre 2000. This bought us a little time, but not much. After tipping point is reached, this less than average incoming energy will help drop us into a new ice age even faster.

How long does it go on warming? Tipping point is the turning point for global warming. At that point it starts getting colder. We will still have a huge amount of Carbon Dioxide and Methane in the atmosphere and this will continue to cause warming over tropical and maritime regions. Over the continents we will see regions that are warmer such as the west coast and Alaska, but towards the northeastern portion of North America we will see drastic cooling. This disparity of temperatures over the continents will cause the severity of weather to increase. The American southwest will see a great increase in rain as will southern California. Central California sees continued drought.

In spite of all the experts talking about continued warming, in 2010 I estimated the tipping point for the next ice age will be 2020 (the next sunspot minimum). I stand fast to this forecast and I expect the summers of both 2019 and 2020 to be very warm and the winter of 2020 / 2021 to be the coldest and longest winter we've seen in 100 years. This has disastrous effects on the climate of the world; especially for North America, Europe, and Asia. After 2020 summers gradually become shorter and northeastern North America sees year round patches of ice, these get larger every succeeding year.

An argument concerning the Sun Spot minimum is simply that we have seen this happen 200 years ago and it warmed up afterwards. The big difference today is during the Dalton Minimum of the 1800's we had a very cold Arctic covered with ice and a strong well organized Jet Stream. That is not the case today as the Arctic ice is rapidly melting and our Jet Stream is in disorder. We have a double whammy coming our way with the Arctic melting and an impending period of weakened solar output.

The right wingers now go batshit crazy and say bull. I say read on.

Imagine growing seasons getting shorter and shorter every year until the world stops producing enough food to support the 9 billion hungry mouths. It is going to get real ugly when ice starts building up and not melting in the summer. This horrifying event occurs within 5 years after the Tipping Point of 2020 (if not in 2021 itself). Central to eastern Canada and the northeastern US is going to get weather that is comparable to the 1880's when Krakatoa erupted and caused a 5 year volcanic winter. Within a decade after this event we will see wars and migrations that change the geopolitical and economic balance of the world.

This was my vision in 2008 when I first wrote Tipping Point and the fact that the sun is going into a solar minimum is making the possibility of it occurring even higher. Sadly, when I first wrote Tipping Point I was unable to get anyone to discuss the possibility of an impending ice age and I expect pretty much the same reception with this re-written version. Such is life.

Even the possibility of a new ice age kicking off is one that should be discussed by governments and citizenry of all nations. If it does occur, it's going to affect us all.

As far as what we can do to stop the world from continuing to warm until the Tipping Point is reached and sliding into an ice age? Nothing, this is an event already programmed to happen long before the rise of civilization. We have just moved up the clock a thousand years as we can't or wouldn't do a darn thing to prevent it from happening. Changing it is nearly impossible, at least not until ice starts building and everyone understands what is happening. Our exploding population and ever increasing reliance on carbon based fuels have pushed up the clock. I don't see any significant changes occurring in government policies over the next 10 years so we are basically on a bus going downhill and our brakes went out long ago. We can be pretty sure of one thing, intense weather patterns will continue as climate is changing; it is just going to get worse and almost nothing we can do will make any difference at all.

Today mankind faces its greatest challenges since World War II, with many economies around the world tottering on ruin, much of our work force is suffering and the darn climate is changing on us. In spite of these critically pressing issues we have a government controlled by elitist groups that refuses to make sound decisions that would help not only this generation, but our children's generations. They only seem to want to get re-elected and wage senseless expensive wars that we will never be able to pay for.

Our illustrious Congress and Executive branches are in effect puppets controlled by any special interest group that supports their re-election. This includes the banking, insurance, oil, farming, logging, plastics, medical, and prison industries that don't make as big a profit if they don't get their way. One instance of this is the demonization of cannabis, if it were legalized they would lose billions a year. Instead they all lobby our esteemed lawmakers, giving them millions for reelection while the country loses 60 billion a year in its prosecution. Money that could be spent more wisely or not spent at all. Thousands of materials can be made from hemp which is why DuPont lobbied to get it made illegal to begin with, it was a threat to his chemical business and a new product called nylon. Of course new industries would arise to replace these, but they don't have lobbyist in Congress to pay off our dirty politicians.

Indecision, denial, and outright lies are the methods the elite politicians use along with spreading fear or hate of any group that opposes them. No one party is infected with this greed and pursuit of absolute power more than another, they all are. It's feeding time at the pig farm and we are the corn.

Our industrial base has largely been shipped overseas and our once proud country reduced to a mere shadow of what it was just a generation ago. Why is this? Because companies make more money when they use cheap labor and avoid strict environmental laws. They sent the factories overseas where this is not a problem, in the process they are laying waste to vast regions of the world and this bill keeps pilling up. It is a powder keg and at some point the fuse will be lit, if it is not already burning, we can see it smoldering in conflicts around the world. What happens when companies start leaving China in masses, when the cash cow (the US) dries up? War is inevitable.

We live in a world of change. I do hope that I am wrong.

Mike Little,

Fairfield Tx, March 2015

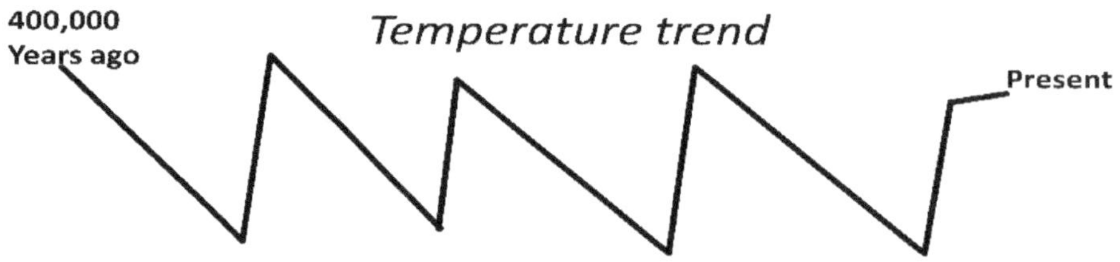

CHAPTER 1

GLOBAL WARMING AND CLIMATE CHANGES

Our environment, geography and time have all played huge roles in shaping us into the adaptable species we are today. We are definitely hard to kill, even in the darkest disaster enough of us survived to carry on the species through trials of fire, flood, and ice that killed off many ferocious top level predators that were simply less adaptable. I discuss many of these disasters, why they happened, and their effects on man and our journey through time.

I have always firmly believed that if you couldn't explain a complicated concept to a simple layman that you were either a pompous ass or had no idea what you were talking about and in either case should just shut the hell up. In that vein in Tipping Point I discuss events over the last 100,000 years that have shaped mankind, using simple non-technical terms. The same way I've always talked about technical matters to layman.

The laws of physics that govern climate are as applicable in the past as they are today and I have applied forecasting techniques I used as a Navy weather forecaster to better understand our past environment. The world is typically a much colder place than we see today, only 24,000 of the last 400,000 years has it been as warm as today and 10,000 of those years were in the current warm era that we live in. We know repeating ice ages occurred, but why? And why did it warm up? This is what I'm trying to get out to the world, a forecast of the weather that humanity will face in the very near future.

Climatology is the study of the past so we learn what variations in weather the future may bring and the past is painting a sad story for us. We near a tipping point; a time when global warming reaches a crescendo, abruptly stops warming, turns and becomes a cooling trend lasting for many tens of thousands of years. It is a repeating cycle (every 100,000 years) that has happened like clockwork four times in the last 400,000 years. The magic key to the start of this cooling trend is the melting of the Arctic Ocean Ice pack (it's the only one that makes any sense), an event happening today. It breaks the Jet Stream and (climatically speaking) all hell breaks loose.

All my instincts tell me once the tipping point occurs climate rapidly changes to one much different than today, at least in northern and eastern regions of our continents. It's as if God left the freezer door open for the last 10,000 years and is finally closing it. We certainly will not get a mile of glaciers in a year or two as this process takes hundreds of years or more (depending on how close the water source is). The problem is ice itself is disruptive to human society, even an inch that remains locked in place plays havoc over the region it covers.

Both man's past and future are closely tied to the climate of the world and for 10,000 years man has basked in the longest warm period in half a million years; this warm period has brought prosperity and enabled us to develop from a time when most people were hungry, to a time when most (but not all) are full. This time of warmth will not last, it will end and soon.

The earth is our space ship carrying us through time and space. Even with all the strife and hardship in our world it is still an absolutely glorious and incredible place, full of surprises and pleasures of all sorts. It is amazing that we live everywhere from the jungles of Brazil to the ice fields of the north and every climate in between, each with different languages and cultures. We even look different. In a very real sense these differences are a result of the stress of climate and the vast distances between developing societies over thousands of years of time allowing human societies to develop in so many different ways. How wonderful it is to be alive. Climate and time have worked together to make us who we are.

Seven years ago I had no intention of writing a book, but the more I studied graphs derived from Ice Core data from Greenland glaciers I began to understand how close we are to a new Ice Age and thought about the consequences of such an event, a disaster of such magnitude that even the threat is such that mankind should prepare or at least talk about it. This dark and foreboding vision compelled me to write my forecast into a book.

Imagine a time when Canadians and American from the northeast are jumping the border into Mexico and they are not welcome. But they go anyway and go armed as they have nothing to lose, already having left a home and life covered in ice. Millions lose everything. A time when war and hunger are the norm and nuclear weapons are used on every continent - more than once.

I do hope that I am wrong.

Ice

The world of 75,000 years ago was much different than today, it had been 53,000 years since the start of the Ice Age and millions of square miles of water had evaporated from the world's oceans, this moisture feeding growing glaciers on our continents. Vast ice fields covered many areas of the northern hemisphere with portions of the ice growing several miles thick. Global sea levels dropped 400 feet and many areas that are today covered in ocean were then above sea level. The very geography of the world was much different.

This fall in sea levels created a Garden of Eden in the Aegean Sea near present day Greece. The Mediterranean Sea needs an inflow of water from the Atlantic or else it slowly evaporates away. As worldwide sea levels dropped the inflow of water gradually slowed in the Straits of Gibraltar (which is only 930 feet deep) and the Mediterranean Sea began evaporating more water that it gained. The Mediterranean water level gradually lowered and split the Mediterranean Sea into three smaller seas, with water levels in the eastern Aegean Sea eventually dropping many hundreds of feet to over a thousand. These gradual sea level drops in the Aegean caused underwater sea mounts to slowly rise out of the receding sea, becoming islands. These were free of the predators of the day and became both a paradise and a cradle for mankind. Many of our ancestors lived there for tens of thousands of years until the ice started melting and took it away. This was Eden.

After the last warm maximum 128,000 years ago, the world gradually got colder and colder and after 12,000 years of glacier building the ice forced man from northern regions into a narrow ecosphere along the world's equator, then sequential eruptions of two super volcanoes nearly wiped us out. The first started erupting 116,000 years ago and was a long series of eruptions of the Yellowstone caldera volcano that dropped the world into an 8,000 year volcanic winter that forced many of the survivors to head south and live near the equator.

The shore line of the Indian Ocean became home to thousands of fishing communities, then after 25,000 years of prosperity came new disasters: first huge earthquakes, and hours later monster Tsunami (much larger than the 2004 Tsunami - but generated by the same fault line) which roared across the Indian Ocean and squashed these thousands of coastal villages. Then came the climax as the supremely violent eruption of Super Volcano Toba in Malaysia occurred, effectively taking away the sun for months and destroying the world's Ozone layer.

Garden of Eden

The few who survived Toba were sheltered in locations with continuous thick cloud cover such as the eastern Mediterranean in what was then a vast and deep canyon (where the Aegean Sea is now located), the Aegean Canyon. The southern side of the Aegean Canyon had the largest waterfall in the world where the gaping mouth of the Nile River poured millions of gallons of water a second over a shear drop into the canyon, moisture bellowing up from the falls created a thick maritime cloud layer that stretched across the entire Aegean, shielding the basins floor from most of the dangerous ultraviolet rays of the sun after the eruption of Toba destroyed the worlds Ozone layer.

This was the Garden of Eden where remnants of the human race were able to survive, a place now under many hundreds of feet of sea water. We had 60,000 years to enjoy this Eden before climate change took it away when the ice melted at the end of the Ice Age and covered this area in water 10,000 years ago.

Why do Ice Ages occur and why do they end?

Scientist just don't know.

It's a mystery to them.

It is the period of maximum warming that propels the world into the next Ice Age, happening when ice in the Arctic melts and the world quickly plunges into another Ice Age cycle. As shown by the Ice Record, the world has done this four times in the last 400,000 years. It seems as if being in an Ice Age is normal.

What happens when it warms up?

Does it just get hotter till we all melt?

I don't think so.

It is a mystery to scientists as to why Ice Ages begin or why they end. As a retired Navy Weather Forecaster I have experience forecasting both meteorological and oceanographic parameters and used that knowledge to try and determine why Ice Ages begin and end. I studied Ice Record data taken from Ice Cores drilled in both the Greenland and Antarctica glaciers; analyzing over 400,000 years of temperature data, looking at trends and variations in temperature for clues.

After several years of study the Ice Record began to scare me, my weather spider sense started screaming that the world is on the verge of a change in climate. In a vain attempt to disprove myself I studied every aspect of the Arctic in search of the trigger that starts Ice Age's, looking at: ocean straits, currents, surface winds, Jet Stream's, air masses, and the overall ocean temperature structure. Studying how major eruptions result in global cooling and gases such as Carbon Dioxide and Methane gas cause warming. Looking into how ice, time, and geographic barriers affected man and his growth or lack of. How volcanic eruptions almost exterminated man and how many cultures have legends that tell of major floods (our ancestors were trying to warn us), these occurred as the glaciers melted over several thousands of years.

I also realized that the ice record shows that man has been modifying weather ever since we began clearing forest by fire for farming 10,000 years ago. Its why climate has been moderate, all the smoke.

Global Warming

Global Warming is a sensitive topic and weather extremes of every sort are continually making the news: blizzards, record snowfalls, record highs and lows, droughts, floods in North Dakota, hurricanes in places that have not seen them for generations, and fast rising rain swollen rivers flooding small communities. The list goes on and on. Something is happening and it's happening today.

Some politicians claim that Global Warming is a myth and any legislation at all is just an insufferable burden on their constituents. Others claim that warming is an unfair burden on third world countries who will suffer much of the brunt of sea level rises and are least able to afford it. I think politicians just like to hear themselves talk, as they do very little of anything else. Just worthless bastards.

Yes it is warming, but it did not start this century. This long term warming trend has been going on for thousands of years. Examination of data from Ice Cores have given us a window into many climatic variables of the last 800,000 years, including: temperature, Carbon Dioxide levels, Methane gas levels, and the amount of dust in the atmosphere. The most important of these variables is temperature which gives us a virtual thermometer into the world's climate for thousands of years into the past.

This Ice Record data reveals some interesting information such as:

- The last time the world was this warm was 128,000 years ago (and it was a lot warmer), then it started cooling and kept cooling for 110,000 years.
- The ice melted between 17,000 to 10,000 years ago.
- For the last 10,000 years temperatures have been moderate and remained fairly steady.
- Cold periods we have seen in the last 1,000 years are minor variations compared to dramatic drops after previous tipping points were reached, these were more severe and long lasting.
- That long term warming and cooling is a long term cycle that has repeated four times in the last 400,000 years in a clear saw tooth pattern.
- The current maximum is more moderate than previous maximums and lasted far longer than any previous maximum.

The world is at the tail end of a long warming period and the hottest part of the cycle is yet to come. It is a warming trend that has been going on for 17,000 years. Almost nothing we can do will cause the world to stop getting warmer, except maybe switch back to burning masses of coal and wood. But that isn't going to happen.

Whatever we do, it is going to get warmer

For the time being until the Arctic Ice fully melts, it is going to get warmer and oceans will rise (at least a foot over the next 40 years). This gradual increase in heat in the world's atmosphere and oceans is going to cause a number of problems:

- The Gulf Stream in the Atlantic (and the Kuroshio in the Pacific) both carry more warm water further northward; increasing both temperature and volume of these northward flowing currents; and the number and intensity of hurricanes (typhoons in the east).
- Creates a greater temperature difference between maritime air masses and continental air masses, giving us much stronger frontal systems. Weather is more intense.
- Areas not normally affected by tropical storms see an increase in hurricane activity: Maine, New York, Washington DC and New England all see more hurricanes and associated storm surges.
- Northern offshoots the Gulf Stream and Kuroshio currents increase in volume and flow, getting warmer and deeper, allowing more warm water to move into the Arctic and hastening the melting of the Arctic Ice pack.
- Other areas have an general increase in precipitation; with more intense rain squalls, severe thunderstorms, and more snow in the winter. These storms often cause localized flooding for communities with inadequate drainage.
- Aircraft see more incidents of turbulence and icing, landing incidents also increase.
- Our infrastructure is at risk, roads, bridges, water and sewer systems all experience larger swings of temperature. Freezing break more pipes, lots of them underground especially in regions that don't currently see much cold weather. The pipes are not buried deep enough in many areas.

- Tornado activity is tied to the jet stream and becomes erratic, decreases in areas that normally has them (due to a weaker jet stream), may occur in areas that don't normally get them. In winter 2015 the number of tornadoes is far below normal.
- Areas on the west coast of continents that traditionally have cold winters, get warmer weather, areas on the west coast favorable to grow wine move further northward.
- On the eastern side of continents storms are more intense with more precipitation and larger differences in temperature; bringing more ice and snow in the winter, heavier rains in the summer.

Why is the world warming? It is a complicated question and we may never know the full answer. Many things affect global warming, some culprits:

- **Carbon Dioxide** is building up in the atmosphere and preventing more (infrared) long wave radiation from escaping at night (the Greenhouse Effect). In the long term climatology has shown us that an increase in temperature precedes increases in Carbon Dioxide. Temperature seems to pull the cart full of Carbon Dioxide (more plant growth). We really don't have enough information to pin global warming on Carbon Dioxide, it's not easy and the Carbon Dioxide in the atmosphere is more due to natural sources than caused by man, but we are pumping a lot into it. This overall increase in Carbon Dioxide is a natural trend that's been going on for 17,000 years.
- **Methane** is building up quite rapidly. All the drilling in the shale formations around the globe is releasing a large amount of Methane, along with the Methane being released from the thawing tundra's in the far north. What methane the oil companies can't sell is being burned off, throwing 25 million tons of carbon dioxide a year into the atmosphere in the US alone.
- A **decrease in reflectivity** of the globe allows more heat to reach the earth's surface. Shifting from coal and wood burning to the use of hydropower and petroleum products has resulted in less smog in the atmosphere. Less albedo and more heat from the sun can hit the ground and is absorbed by the land surface, this in turn heats up the lower atmosphere.
- A **decrease in land and ocean areas covered by ice and snow** affects global warming to a HUGE degree, as it takes 80 times the energy to melt 1 gram of ice than it does to raise 1 gram

of water one degree; this means that after an area of water is no longer covered by ice, the water tends to warm up 80 times faster than it did when it was covered by ice.

Who is to blame for global warming?

Did Bush, Obama or anyone in American politics create policies that have directly caused global warming? What a wild thought and it assumes that we have one politician in America that actually is capable of doing something besides maintain the status quo, whine, and complain about their political opponents, that they are "Not Fair". Just a bunch of do nothing crybabies whose only goals in life is to be take money from lobbyist and anyone else with a bribe, and to be re-elected. They are all traitors to the people.

Can one person's day to day activities have the ability to affect global weather? This is difficult to believe as climate is generally a case of effects canceling each other out; for instance if a whole bunch of people cause a large amount of global cooling (or is it warming) by buying electricity generated by coal; another big bunch of people will be doing things that cause global warming (or is it cooling) like passing clean air acts, or driving electric cars. **We know politicians have done little to affect climate, there's no money in it.** Henry Ford inventing the assembly line method was responsible for some climate change – but I am not sure if it was heating or cooling (likely was cooling).

Let's blame the Cows

Worldwide we have over 1.5 billion cows who each produce over 100 gallons of methane gas a day from belching and farting. This generates over 100 metric tons of methane a year. Crap, Congress can do that in a day.

The Blame Game

I hope everyone quits whining and pointing blame about something we cannot control, it's like complaining "the sun is too bright and OHHHH it's HOT too", just useless whining. We can assume that it is warming and can do little to change that fact, the time for placing blame and change is long past. Global warming and Carbon Dioxide emissions are a marketing ploy used to make money and for politicians to be elected. Can we stop the warming trend by cutting back on Carbon Dioxide emissions? This is unlikely as it is doubtful that anything mankind can do will rein in the runaway increase in global temperatures; it is simply going to happen, it is a trend that's been going on for 17,000 years and very unlikely that anyone or even everyone together can stop it. It is simply the natural order of events.

Even if all the world's governments got together and made real and significant changes in Carbon Dioxide emissions, it's possible that Carbon Dioxide is a result and not a cause of global warming. We just know so little and at the end of the day it is going to be a case of knowing "too little, too late".

Besides if a significant reduction in Carbon Dioxide emissions did occur, while it may decrease the greenhouse effect it is also accompanied by a decrease in the amount of smog in the atmosphere leading to a corresponding significant reduction in the amount of sunlight getting reflected (by smog) back into space. So once again, good intentions may have entirely opposite consequences than intended, such as causing more surface heating.

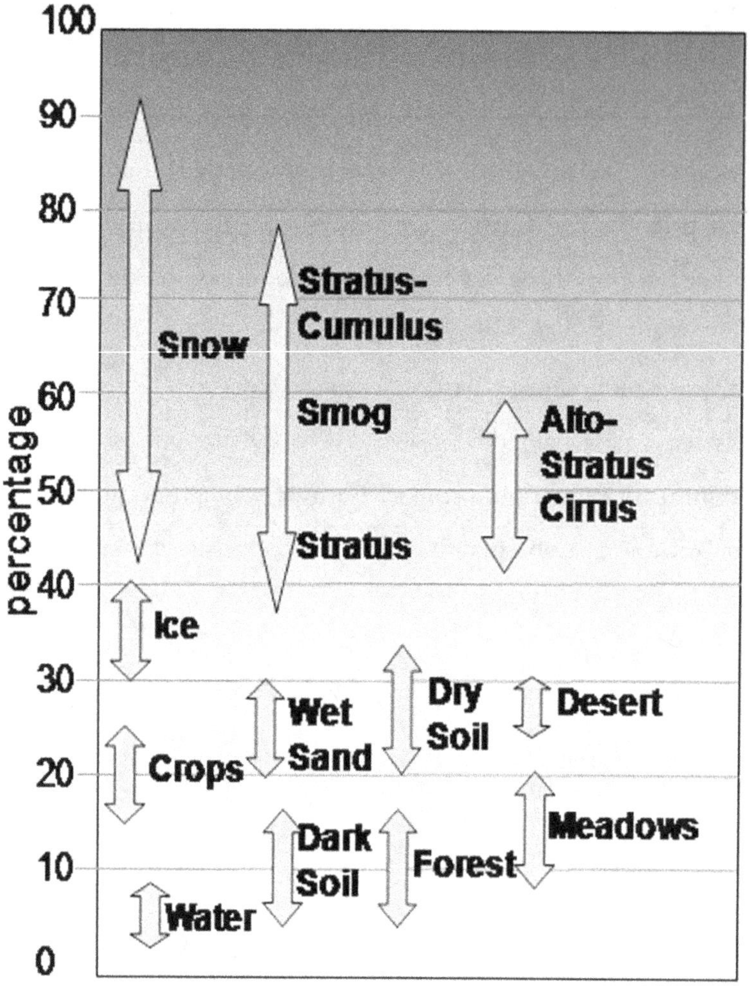

Graphic 1. Albedo of various surfaces

Albedo

Albedo is the percent of reflectivity of a surface gauged against a total of radiation being received by said surface.

An example is how much sunlight is reflected back into space from the top of ice or thick clouds compared to how much total sunlight is received.

In the 1800's the sun went into a period of minimum sunspots and the overall output of energy from the sun dropped just a fraction. This minor decrease in energy was enough to affect the climate of the world to a huge degree (granted we also saw a lot of volcanic activity in this period). Many years in the 1800's summer simply did not come. This tells me the earth's climate is sensitive to decreases in albedo.

Albedo

Smoke and resulting smog have a greater effect on the worlds Albedo and climate than anyone has considered. Consider Graphic 1, stratus (stratified layer – like a blanket) reflects 50 percent of the incoming radiation. Pollution in its finest form (read smog here) acts very much like stratus, but has more and finer particulates to reflect sunlight and does a better job reflecting 60 percent of the incoming radiation. Not much of a change, but more than enough to change history.

During the last 10,000 years man has burned a lot of wood, coal, peat moss and other fuels; releasing enough particulates to do a great job of polluting both Europe and Asia. Prevailing winds carried this pollution east and affected the Albedo of the world's largest land mass. This smog increased the overall reflectivity of both Europe and Asia and instead of causing warming it moderated the temperatures of the world, resulting in a milder and far longer lasting warm period than would have occurred without man's intervention, this is clearly shown by Graphic 3. Every previous heating period (leading up to the four temperature maximums over 400,000 years) has been quick and decisive, while this time it just stopped warming as soon as we began polluting the air 10,000 years ago.

I argue that man has been modifying the climate for thousands of years, without man's use of fire and polluting the atmosphere we would be deep in the next Ice Age.

Inversion

In a normal atmosphere, temperatures decrease with height.

An inversion is an atmospheric condition where temperatures increase with height.

This might occur over areas of cold water and ice where the surface cools the air above.

The Inversion cycle and Smog

An inversion occurs when you have warm air over a cold surface, the air near the surface cools, sinks and becomes more stable while the air above is still warm. An inversion creates a thick layer of clouds covering the surface that acts like an insulating blanket. If smoke is also injected into the inversion layer it is trapped and covers wide areas like a thick dirty brown blanket. Smog lays in a layer across terrain and is black and opaque, causing heating to occur at the top of the pollution layer. This heating increases the temperature difference between the cool surface and the warm top of the inversion; increasing the stability of the captive air mass, making it harder for the inversion to break and for the smog to be released into the atmosphere above. Smoke increases Albedo at the inversion layer, reflecting more sunlight back into space during the day and increasing cooling at night from the top of the smog layer (cooling is the effect).

Fire is a tool man has used for many generations, it's had widespread use for at least ten thousand years as seen in the Ice Core records and likely long before. It may have been in an oil rich land in the Middle East that we first found fire. Where natural seeps of oil and gas occur (and many places elsewhere) that have been burning for thousands of years, probably ignited by static electricity. These are natural eternal flames and certainly visible at night, they may have been safe havens that ancient man huddled around at night. We have at least one such seep in the US that has been burning for hundreds of years, we don't know how long. In any case such a seep would have been a perfect place for man to learn to utilize fire.

After learning to use fire, we quickly climbed out of the Stone Age. Consider that man began clearing land for farming by burning virgin forests thousands of years ago. Plus the mandatory nightly use of fire to simply hold back the evils of darkness and the ever increasing use of fire in industries such as: pottery, copper, brass and bronze; resulting in a lot of material being burned.

Looking at long term temperature trends (Graphic 2 and Graphic 3) derived from the Ice Core data, it is clear mankind's use of fire has changed the climate of the world. But not as expected, not as warming; instead by keeping the planet cooler than expected in a warming period. Previous warming maximums like the one that happened 128,000 years ago were rapid and short. The current maximum has lasted for nearly 10,000 years, almost as long as the previous three maximum's combined.

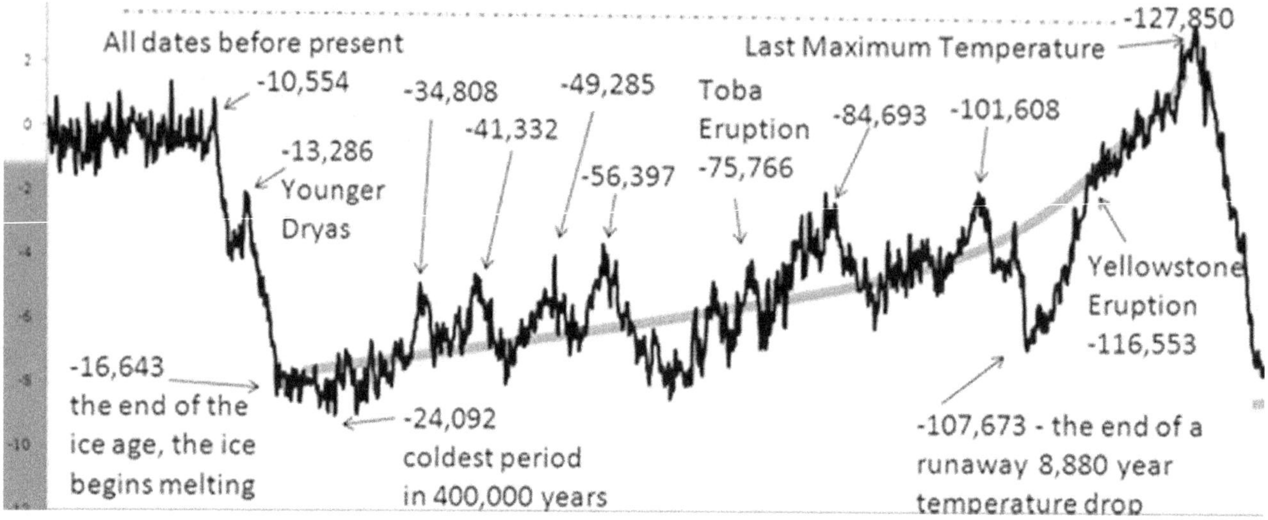

Graphic 2. Long term temperature trend from Ice Core Data (years ago verses changes in temperature in Celsius) today is on the far left side (Data from Petit, J.R., et al., 2001)

For 110,000 years the world got colder and colder, only 17,000 years ago did it start warming. The gray thick line on Graphic 2 shows the general long term trend, this period of continued cooling must have a single cause or we would see many more random fluctuations of temperature. Instead we have a single event that lasts for 110,000 years that has repeated four times in the last 400,000 years (see Graphic 3). The only cause that makes any sense and lasts so long is the gradual long term evaporation of the world's oceans and the creation of glaciers on the continents..

Methane Hydrate

Methane Hydrate looks like snow and most every open space within the hydrates icy crystalline structure is packed tightly with Methane gas. Hydrates are conditionally stable as long as temperatures are kept below 50 degrees Fahrenheit and pressure above 500 pounds per square inch. These hydrates are important because a huge amount of Methane trapped in its crystalline matrix, when it melts it degases into water and Methane. This occurs if either temperature increases or pressure is decreased then hydrates may release their captive gases, this can occur spontaneously over wide areas of the ocean floor and literally trillions of square feet of this substance lay on world's oceans continental shelves. A time bomb, waiting to go off.

So now the question becomes, why do temperatures vary above or below the general trend?

Carbon Dioxide Gas (large long term changes) – causes large scale temperature maximums. We are at the tail end of one of these temperature maximums.

Methane Gas (short term changes) – during the last Ice Age, six temperature spikes go above the long term trend on Graphic 2 and each spike shows approximately 7,000 year periods of warming. This is a result of massive sea level drops that caused huge areas of Methane Gas trapped in Methane Hydrates on the world's continental shelves to become unstable and degas, the result of warming oceans and decreasing sea levels.

In our major ocean currents we have warm and cold core eddies that naturally form (around the major currents of the world) and then drift slowly along with these currents. In the ice age as these warm core eddies moved over conditionally stable methane hydrate deposits we saw huge degassing events. Today we have seen some minor events over these massive Hydrate deposits, plus events in the cold tundra areas such as multiple massive blackened holes (hundreds of feet across) that have been found in Siberia, these are formed as Hydrates explosively degas.

Now fast forward to today and the question rapidly becomes: **"Does the greenhouse effect cause more warming than the amount of smog and smoke in the atmosphere causes cooling?"**

This is an interesting question about Smog and the Greenhouse effect, when these two combine they generally negate one other and do little to increase the average worldwide temperature as they balance each other out (as smog did over the last 10,000 years). But in the last hundred years man has changed the equation, shifting away from dirty wood and coal to a much cleaner sources of energy such as oil and hydroelectric power; much cleaner than wood or coal and less pollution. This created the combination of decreased smog and pollution (and decreased Albedo) that resulted in an increase in surface heating over the major land masses and an increase in Carbon Dioxide that resulted in the earth retaining more heat at night. Did this cause the resulting rise in temperature? That is my guess.

Just a hundred years ago our air was polluted from wood fires used the world over for industry, heating, and cooking; and for most of this century wood and coal continued to pollute the air. It was only in the 1960's that laws were passed that largely cleared the air over much of the Northern Hemisphere. Before this time the additional soot and smoke in the atmosphere prevented a small percentage of heat from even reaching the ground. After stricter environmental laws were passed, the air cleared and the world started heating up very rapidly.

It is possible that today's heating
may be caused not so much by an excess of greenhouse gases
But by the efficient use of fuels that cleared the air
allowing more of the suns energy to hit the ground.

Both technology and environmentalists are to blame.

CHAPTER 2

THE ICE AGE TRIGGER

The search was clear, why do multiple Ice Ages occur in a cyclic pattern, four times in 400,000 years. What logical reason explains this strange behavior? What explains this cyclic pattern and is also meteorologically sound reasoning? I assumed that the Arctic Ocean was free of ice for most if not all of the past Ice Ages, as this has been shown to be the case from cores drilled in the Arctic sea floor.

Let's go over what does NOT cause it

The average length of an Ice Age is 100,000 years, this pretty much rules out cyclic variations in the suns solar output and variations of the earth's rotational orbit, as both of these events happen with much greater frequency, no clear correlation exist with these and the patterns exhibited in the ice record. It has to be something else.

My quest was simple.
What variations of meteorological and oceanographic conditions existed 128,000 years ago that worked together in such a manner to cause ice to form for tens of thousands of years and not thaw?

To better understand the overall climate of the Arctic I studied many parameters such as: how the position of land masses influence weather patterns, looked at surface and subsurface ocean currents, and wondered how much the depth limited straits leading into the Arctic Ocean plays in restricting the flow of both warm and cold currents between the Atlantic/Pacific and the Arctic Oceans. I then looked at how changes in currents and atmospheric conditions in the Arctic affect the overall climate of the world.

Most importantly with my background in Meteorology and Oceanography I thought about how all these work together.

I conclude that Ice Ages are a result of a domino effect:

1. The first domino is heating causes the Arctic ice pack to melt.
2. The second domino is warming in the Arctic Ocean disrupts the world's Jet Streams.
3. The third domino is a weakened Jet Stream is unable to push air masses out of source regions in North America, Greenland, Europe, and Asia.
4. The last domino is these air masses stagnate over North America, Greenland, Europe, and Asia.
5. Glaciers begin the long process of forming under these stagnated air masses.

In the belly of the beast is the answer; in an Ice Age air over glaciers is intensely cold, this is an air mass sitting on ice and creates intensely cold conditions. These Ice Age air masses have three distinct characteristics: first they are extremely cold (at least minus 40 degrees Fahrenheit), second they are stationary and remain in one place for the best part of 100,000 years, lastly they are extremely dense at the surface creating extreme high surface pressures with an equally strong area of low pressure aloft (nature loves balance, a High at the surface equals a Low aloft, except with warm maritime Highs such as the Pacific High). Even a hurricane changes wind direction aloft, around 20 to 25 thousand feet it switches direction, what goes in the bottom comes out the top (or the reverse). The chimney effect.

Air masses sit over source regions and generally start migrating into areas due to differences in the speed of upper level jet stream (called jet maximums), these kick air masses out of their source regions and they start moving into other regions. This generally creates frontal systems and a lot of the weather as we know it today.

Stationary upper level low pressure systems above cold glaciers create strong cyclonic winds aloft (20,000 to 40,000 feet.), spinning eternally around these upper level lows. They act as blocks and are instrumental in weakening and breaking up the Jet Stream. This lack of a healthy Jet Stream prevents these air masses over the glaciers moving into other areas and these remain locked in place, getting colder and colder for thousands of years.

Conversely, it is dropping sea levels that eventually end Ice Ages.

Icing

Icing is a severe threat to aircraft of all kinds, as ice is heavy.

It is especially a threat on takeoff.

Propeller driven aircraft and helicopters are more affected than jet aircraft as these craft are limited in how high they fly and often fly in clouds that may cause icing.

Look out when temperatures are between 32 degrees Fahrenheit to 10 degrees Fahrenheit.

CHAPTER 3

MAN AND CLIMATE

Climate has played a crucial role in humanities growth from the Stone Age to the present. Isolation caused by a combination of severe climate in the Ice Age's and geographic barriers influenced the very growth of societies. The vast varieties of cultures around the world today are a reflection of that isolation and of mankind reacting and adjusting to both climate and geography. The color of our skin, the vast varieties of languages, and every difference between all the worlds cultures are all reflections of the great rifts caused by climate and the geographical obstacles between growing human societies. Climate, geography, and time have indeed made us who we are.

We also know from experience that worldwide temperatures drop during volcanic eruptions due to the ash and other pollutants in the atmosphere preventing a larger portion of sunlight from reaching the ground, this is increased Albedo (see Graphic 1).

In April 1815, the eruption of the Malaysian Tamboro Volcano (ejecting 150 cubic kilometers of ash) and dropped temperatures so badly that the year later (1816) is known as the year without summer. It was during this dreary year that Mary Shelly wrote Frankenstein.

The eruption of the Philippine Pinatubo volcano (ejecting 3 cubic kilometers of ash) in 1991 also created a noticeable drop in temperatures worldwide. I was 21 miles away and my first child was born this year, she was just a few months old when she was evacuated from Subic Bay Naval Station on board the USS George Washington (a US aircraft carrier), along with her mother. I stayed behind.

Looking at Graphic 2, it is clear that 116,000 years ago a volcanic event occurred (Yellowstone erupted in this time frame) that dropped global temperature in a way that can only be described as a volcanic winter. Temperatures dropped faster than they have anytime in the last 100,000 years, about 1 degree Celsius per 1,000 years. This drop in temperature occurred for 8,000 years straight and created extremely cold conditions that did not happen again until Toba Caldera erupted 30,000 years later (around 75,000 years ago). This later super eruption of Toba was more intense but not as long lasting and nor cause as abrupt a drop in temperatures.

In fact the eruption of Toba, what is classified as the second worst eruption in several million years is rather difficult to pick out on the ice record, only by also looking at the amount of dust in the atmosphere does this event show up. It kicked up a lot of dust that took thousands of years to clear out of the air. This eruption of Toba has been blamed for a near extermination of humanity, but it may very well have been the earlier event or the deadly combination of Yellowstone, Toba and the Ice Age that was so effective at killing off almost 99% of humanity.

At the height of the last Ice Age after ice forced most of humanity out of northern Europe and into a narrow band around the equator; the huge caldera volcano (Toba) in Sumatra erupted, nearly exterminating humanity. Between the stress on society by ice in the north and a major eruption along the equator, the situation became unbearable and resulted in the widespread resulted death of humanity. This was a time when only 1 in 10,000 females survived, scientist call it a DNA bottleneck. As a comparison in a city of 10,000,000, only 1,000 breeding age women survive and in a world of approximately 200,000,000 humans (74,000 years ago) only 10,000 fertile females survives. It was close, real close.

Toba Volcano is located in Malaysia and directly perpendicular from the earthquake area that generated the deadly 2004 Tsunami. Toba has a huge magma chamber (caldron) under a lake created in the last eruption when magma emptied out of the caldera far below, this a hallow depression that is now a lake 60 miles long and 20 miles wide. Today, an island rising in the middle of the lake is a reflection of the mass of magma again building up in the caldron, with pressure pushing up from far below. Heat from the huge amount of friction generated by the 9.2 magnitude earthquake in December 2004 is still slowly percolating upward towards Toba's magma chamber. This additional heat may awaken the sleeping dragon but we have no way of knowing; if it does we will all feel its affects as Toba is a global killer. Only time will tell.

As a comparison to Toba's eruption 75,000 years ago, in 1991 Mount Pinatubo in the Philippines erupted, this was the largest release of magma in the last 100 years and threw over 3 square kilometers of magma far into the atmosphere. 3 square kilometers of magma is a lot of ash (about 1.15 square miles), at 120 pounds per square foot - this equates to forty trillion pounds of ash. I ended up with 8 inches of ash on my home that weighed 80 pounds per every square foot. A lot of homes and businesses had collapsed roofs, the Navy base alone lost almost 500 building (metal building are not good in an ashfall). Toba has a magma chamber that holds more than 3 thousand square kilometers of magma, roughly a thousand times larger than Pinatubo and equates to over forty thousand trillion square meters of highly compressed magma. If during the eruption of Pinatubo a thousand's times the ash had fallen on my home it would have been crushed under the ash long before the eruption was done. As it was during the eruption I was thinking about Pompeii and beginning to panic, had post-traumatic stress for months afterwards.

Our scientists have no idea how a caldera volcano erupts, they deal in only facts and completely unwilling to make suppositions on what they have never seen or measured (a failing I don't share). Caldera volcanoes are definitely not your typical cone volcano, instead they are usually big lakes with an island right in the middle (like a pimple on the earth's crust). Toba covered areas of India (several thousand miles away) with dozens of feet of ash. It takes a lot of force to throw ash that far, perhaps enough force to launch material into orbit. The dust and smoke Toba lingered in the upper atmosphere for years.

Favorable climate during the last 10,000 years has been critical for man's ability for growth, allowing mankind the luxury of developing from Stone Age technology to the Space Age. Looking at Graphic 2 this shows a temperature trace derived from ice cores drilled in glaciers in Greenland and gives a contiguous temperature profile for the last Ice Age cycle (the cores show 400,000 of temperature, graphic 2 focuses on the last ice age). Graphic 2 clearly shows that beginning 10,000 years ago the northern hemisphere enjoying moderate climate. This was a reprieve from the ice that covered much of the continents for the previous 110,000 years. This moderate climate allowed mankind to mature into the global society we have today. This warmth has been a godsend.

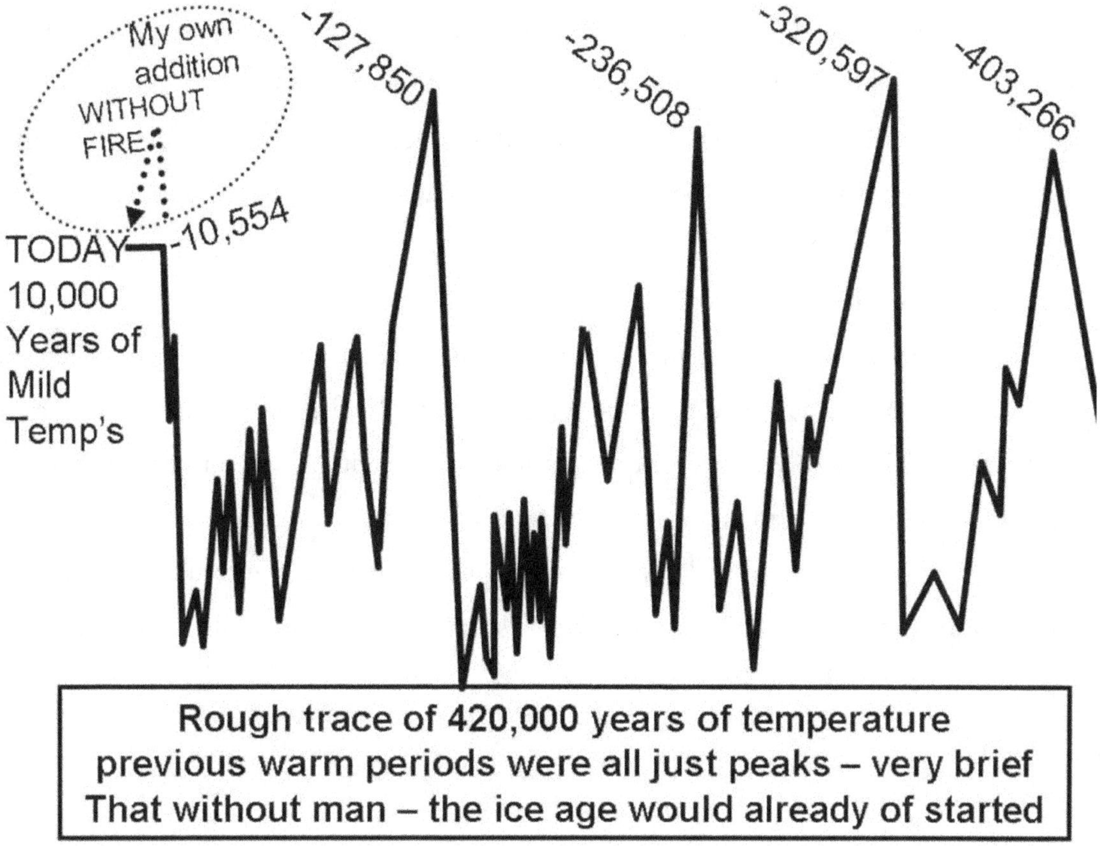

Graphic 3. Temperature trace of data from Ice Cores covering 400,000 years

Graphic 3 shows an expanded view of the last 400,000 years of temperature traces and shows the relationship between the period of maximum heating and the start of the last Ice Age, The last warm period maximum was brief and followed immediately by a long and gradual cooling trend. Graphic 3 also shows that man has been modifying weather far longer than previously thought (contrary to popular belief Al Gore and his low mileage fleets of SUV's were not the start of it, it was an ancestor of his). For 10,000 years we have been burning anything we could and definitely modified the climate. It appears that climate is sensitive and easily modified. In the upper left of Graphic 3, I have drawn in dashes what the temperature trends should have been if Al Gore's ancestor had not tamed fire (he probably traded a skunk hide for the flaming ember, then said he created fire). If trends had been similar to four previous warm periods we would be deep into the next ice age.

When I first started studying data from the Ice Cores that combined Carbon Dioxide, Methane, Dust and Temperature in data traces covering 400,000 years, I was unable to fathom how these trends made any sense at all. They clearly were the work of a crazed madman and I was unable to understand how they fit together; only after separating the temperature trace from the rest (as shown in Graphic 3) did I began to see a clear pattern that repeated over and over, **a clear saw tooth pattern**.

Why a Saw Tooth pattern?

Many environmentalist and politicians are upset about Carbon Dioxide and associated increased greenhouse effect, global warming, and corresponding rise in sea levels. But we really do not even know if Carbon Dioxide buildup is a result of heating (from increase plant growth) or if heating is a result of Carbon Dioxide buildup. Much attention has been paid to short term temperature increases that have occurred within the last 100 years, while little to no attention given to the trends of the last 400,000 years. All the learned scholars of the world seem to think that warming will go on forever and the planet will eventually just melt down.

I think they are all wrong. Pretty typical of me, always 180 degrees out of step. Should have been a salmon, I like swimming against the flow.

MJL

Some Rules of Thumb - not absolute – but pretty firm:

- Surface cooling results in dense cold air and high surface pressure, above the surface cooling (above 18,000 feet), conditions reverse and become relatively warm (for the height) and Low pressure aloft.
- Warming at the surface results in unstable air and low surface pressure and high pressure aloft (above 18,000 feet).
- Nature and laws of physics govern the dynamic processes that occur in our atmosphere, these laws require that the atmosphere does whatever it takes to stay in balance. It's like a chimney, what goes in the bottom eventually comes out the top.

As a child with my parents I visited Los Angeles in the late sixties and the air was completely choked with smog. The sky was brown and it was impossible to see more than a few blocks, not a very inviting place to live. I was glad to leave. The pollution stretched into Arizona as a gray cloud laying across the width of the Grand Canyon. Today industrialized countries of the world are still doing their best to pollute as much of the world as possible, but just a few decades ago the world was much more polluted; air across much of America was dirty.

A hundred years ago air in Europe was even worse from pollution from all the wood and coal fires used for industry, heating, and cooking. A lesser type of pollution continued to pollute the air till the 1960's. This pollution stretched across a portion of Europe and Asia. The albedo was high during this period and prevented a large portion of the sunlight from hitting the ground. In the day it accumulated at the top of inversion layer and was radiated out to space at night.

Dust and soot from man's fires over the last 10,000 years have modified the world's climate, helping to keep a small portion of the suns energy from reaching the ground. This helped keep the world's climate in balance, preventing it from rapidly warming and melting the Arctic Ice Cap. Only after we started clearing the air by shifting to oil, methane, nuclear, and hydroelectric energy did the earth really start rapidly warming again. Plus prior to 1900 much of the pollution was injected into the atmosphere in Europe and Asia, right where the increased reflectivity had the most impact on the world's largest land mass. Right where it most effectively cools Europe and Asia.

Recent global warming is caused not so much by an excess of greenhouse gases, but by the efficient use of fuels and anti-pollution regulations that have done a great job of cleaning the air, now allowing more of the suns energy to actually hit the ground. I am a liberal and really hate to admit this but it is the damn environmentalists with good intentions that pitched us into the next Ice Age. Certainly not the first time that someone with good intentions have screwed things up. We are Americans after all; it's our birthright to get involved in any situation, then screw it up. Of course we do know what's best for everyone, especially if they don't agree.

The historical temperature trace of Graphic 3 shows 400,000 years of temperature traces from the Greenland Ice Cores, it show several interesting things:

- Four previous periods of maximum heating (except for the most recent one) rapidly peaked and were immediately followed by rapid cooling.
- After maximums, each long term cooling trend continues until the world abruptly warms again and is a sustained, gradual, and long lasting cooling trend; on the average taking 100,000 years.
- Each long term gradual cooling trend has short periods of rapid cooling trends that are a result of volcanic eruptions (these correlate well with known historical eruptions).
- Each long term gradual cooling trend exhibits short periods of rapid warming, the result of huge deposits of Methane Hydrate degassing off the worlds continental shelves in response to large sea level drops and warming oceans. This causes heating due to Methane's greenhouse effect, which is more effective than Carbon Dioxide (anywhere from 33 to 72 times more effective, it's debatable and experts disagree how much).
- Graphic 3 shows classic repeating saw tooth patterns.
- That the last 10,000 years was not at hot as previous warm periods, caused by man's burning of wood and coal and the resulting smoke and smog modifying the world's temperature.

Mankind's polluting of the atmosphere has been instrumental in prolonging the latest warm spell. This increased the amount of reflectivity of the world's atmosphere, especially over the continents where surface heating is most effective. Four previous warming period lasted only a few thousand years, while the current warming trend has lasted 17,000 years (we are WAY overdue for a new ice age).

The only difference seems to be that man used fire extensively for industry, cooking, to clear forests for cultivating the land, and holding back the evils of the night. This pollution helped cool the planet, resulting in the world not warming up as much or as fast as without it.

Along with this deduction came the idea that global warming is the cause of an Ice Age, that this is the earth's method of staying in balance.

The bottom line is that in the end our attempts to clear the air, use our fuels more efficiently, and to create a world that is more livable may actually be responsible for causing the earth to heat faster, and ultimately be responsible for creating a chain of events that cause the earth to drop back into an Ice Age within the next generation.

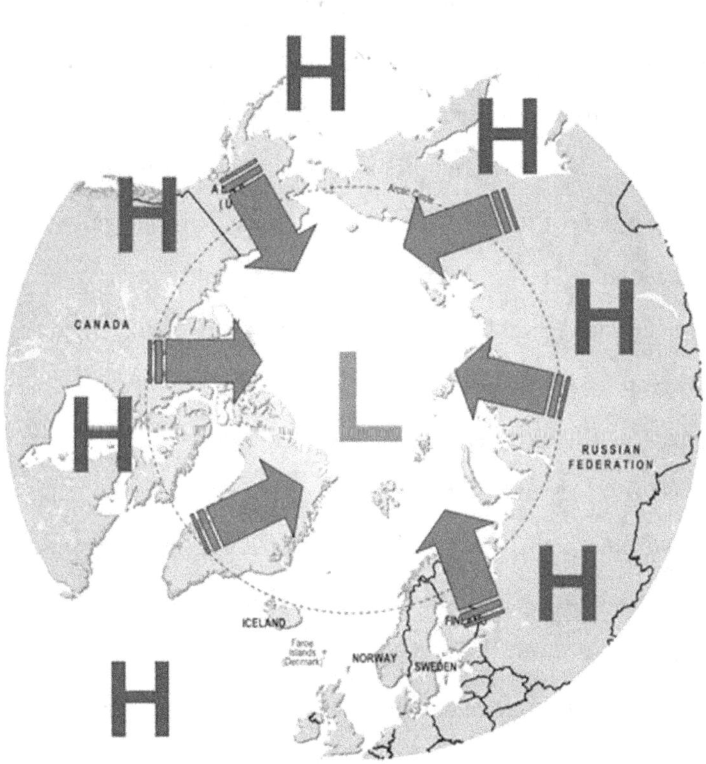

Graphic 4. Surface winds blowing northward with ice free Arctic Ocean map courtesy of http://maps.grida.no/go/graphic/arctic-map-political

Why the Ice Age begins and why it stays

Imagine a time when the ice cover in the Arctic Ocean has melted, when the surface water is too warm and storm tossed for ice to readily form. This leaves an ocean surface area that is warm and moist; this has a huge effect on the weather patterns of both the Arctic and Northern hemisphere.

During an Ice Age:

1. The area of open water in the Arctic is much warmer than the land masses of Canada, Greenland, Iceland, Europe and Eurasia. This creates an area of Low pressure over the warmer water. This low is caused by surface warming and partially because it is an induced Low between strong surface Highs (like a valley between two mountain peaks).
2. Cold air predominates over the northern continents and begins to blow northward towards the Arctic Ocean. This wind flow is shown in Graphic 4.
3. Cold air blows northward goes over warm open water as shown by Graphic 4, this wind evaporates moisture quite rapidly from the ocean and becomes unstable and rises, this moisture and warmth eventually reach great heights in the Arctic (half what it would be in the tropics).
4. In a reversal from the rest of the world's oceans where warmer water overlays cold water.
   In very cold regions such as the Arctic Ocean, very cold water overlays warmer water. Winds blow northward over the Arctic Ocean causing upwelling along the coastlines of the Arctic, this brings up warmer water from below.
5. It is the warming of the atmosphere caused by contact with warm ocean water and the continuous addition of moisture that creates Low pressure on the surface of the Arctic.
6. Cold air and low surface pressures create an upper level high pressure system. This is an upper lever ridge of High pressure that extends from Alaska over the Pole to Iceland. This prevents the Jet Stream from becoming organized and cuts the Jet Stream into a series of weak Jet Segments that rotate around permanent cold surface Highs over the glacier covered continents. A weakened Jet Stream splits into segments that are not strong enough to push the intense cold air masses out of the interior of the continents.
7. Surface cold air masses over the continents are far colder than they are today and permanently located over the glaciers on the continents. The surface warm air mass over the Arctic is also stationary. These air masses remain stationary for tens of thousands of years.

Imagine a day in Eastern Canada or the Northeast US with temperatures far below zero; ice and snow came early that winter and fails to melt. A very cold air mass builds on the surface and an occasional warm moist wind blows over the very cold air, freezing rain falls and accumulates as a layer of ice. This condition of surface freezing and moist air flowing northward from the Gulf and Atlantic continues all winter. When the short summer finally comes, this layer of very cold ice is protected by an inversion layer of stratus and fog shielding the ice from the sun. The inversion layer holds firm and most of the ice stays all summer and the process repeats year after year, with succeeding summers becoming briefer and briefer.

Why do Ice Ages occur at all?

In the eyes of the esteemed scientists of the world, it is a mystery why Ice Ages start or why they end. I question if an experienced weather forecaster ever analyzed why Ice Ages occur or why they last for 100,000 years in a cyclic repeating pattern? A weatherman is certainly a more appropriate choice than a proctologist, geologist, or astrophysicist; even a meteorologist generally has little real world forecasting experience (they generally read what the computer tells they to say, talking heads). Is it really necessary to have these non-forecaster types tells us why a cold air mass develops over continents in the northern hemisphere and lasts for 100,000 years?

Some experts in underwater currents say it is the interruption of high density subsurface currents originating off the ice shelves of Antarctica that cause Ice Ages, others look at the melting Greenland Glacier and see it as a source of fresh water that melts and flows into the sea, then interrupts the flow of ocean currents in and out of the Arctic Ocean. My take is that fresh water is lighter than salt water and Greenland melt water will float on top of heavier salt water (it also depends on temperature). It is complicated and makes my head hurt but if I get it right, the theory is this influx of fresh water into the ocean on either side of Greenland will interrupt either the flow of warmer water into the Arctic or the outflow of cold water. I say it might interrupt these currents for brief periods, but not for thousands of years. Besides glaciers usually release melt water to the sea in great spurts as the melt water finds cracks in the ice, but these cracks refreeze and water pools again. Hence the legends of floods.

My question is simple, do ocean currents cause weather or does weather cause ocean currents? I think the oceanographers may be putting the cart in front of the horse as it is the weather that generally (almost always) cause most currents. It is the winds, dense water created by evaporation, and cooling that create ocean currents. Not the other way around. In either case, melting glaciers cause short term events and the Ice Age last for 100,000 years on the average.

Other revered scientists study the sun and point to long term variances in the sun's energy output as the reason Ice Ages occur. This is actually possible but these events occur much more frequently than every 100,000 years, it really isn't possible for the sun to stop shinning so brightly for 100, 000 years and then do it over and over in a cyclic pattern.

Some scientists with way too much time on their hands look to the alignment patterns of the planets in the solar system and claim the alignment of Jupiter or Saturn cause Ice Ages and "**AHHHHHHH"** we are all going to die on Dec 2012 (or whatever date they pick as the next doomsday). I do realize (to my own huge dismay, crap) that I am a weather doomsday'er. I guess Chicken Little was an ancestor and it runs in my blood, I can't help it.

Other scientists and experts (usually with real bad hair) look light years away towards the center of the galactic core and wonder if gravity waves or other equally far left field effects from a massive black hole at the center of the galactic core causes Ice Ages. The idea is the effect of faraway stellar or interstellar objects a hundred lights years away are able to cause Ice Ages like a huge cosmic death ray. Respectfully, I look at all these highly educated, probably well-funded, and knowledgeable sources as people that are completely out of their minds and really need to get a day job (I have one - thank you). The bottom line is no one to date has given a credible and logical answer that clearly explains how Ice Age's start or why they end. Where is the smoking gun? Where are the dots that connect the cause to the event? Random changes do not cut it, not when it happens in a repeating 100,000 year cycle.

The world's environment is extremely complex, even the largest and most sophisticated of numerical weather programs can only run weather models created by man and these models are generally based on what has happened in the last 50 years. They can handle a steady state or moderate changes in weather, but unable to adjust to periods of time where the weather begins to vary by great extremes.

The old saying is true,
we know what we know, but don't know what we don't know;
AND
About why Ice Ages begin or end,
we don't know much at all.

The face of the planet shows evidence of great events that happened in the past, North America and many other parts of the globe have huge scars where glaciers have cut into the surface of the earth like giant bulldozers, scraping down to the very bedrock. From mankind's limited viewpoint these events occurred thousands of generations ago, so long that we have no memory or records, so long ago that they seem unimportant. Yet from a geological timeframe these events were a blink of an eye in the past.

Scientists have analyzed ice cores taken from glaciers in Greenland and Antarctic, these show 375,000 years of the last 400,000 years (94% of the time) North America and Europe and parts of western Asia have much colder than they are today. It seems as if having continents entombed in ice is normal. Only 24,000 of the last 400,000 years has been warm as today and that is counting the last 10,000 years. What is unusual is for the earth to be as warm as today and to stay that way for long periods (as in the case of the mild weather we have seen for the last 10,000 years).

My contention is simple, mankind is living in a time of great change and what has happened yesterday is not necessarily what is going to happen tomorrow. Change can happen and catastrophes will come, these include randomly spaced huge volcanic eruptions and cyclic Ice Ages.

What is important is that change is GOING to come!
We have to prepare.

Our planet is currently in a period of extreme warming and no matter what we do, it will continue to warm. No amount of crying, tree hugging or legislation by politicians of any kind is going to make any difference at all. No amount of yapping and gloom casting is going to close one power plant in India, China, or America and even if they all close the warming trend will continue.

Warming is a natural trend and will continue despite anything we do; the release of Carbon Dioxide in the last 20, 100 or even 1,000 years has done little to change the course of climatic history. Prior to 1960, man's use of fire and the subsequent polluting of the atmosphere had prolonged the current warm period that has allowed man to develop to the level of culture and technology that we have today.

When the next Ice Age begins, we may actually release masses of Carbon Dioxide or Methane Gas, to enhance the greenhouse effect and try to keep the Northern Hemisphere warm. We might want to stock up.

We have huge problems in the world, and politicians and government should spend more time trying to solve problems and guide our future, rather than determine blame for past mistakes that we can do little about. In that vein none of my theories have anything to do with laying blame for global warming; instead it is my belief that mankind should rejoice as we have enjoyed the longest warming period in the last 400,000 years. For several years I have searched for a meteorologically sound answer as to why Ice Ages begin one day and last for over 100,000 years and have developed a theory on why it does this.

The only answer that makes any sense
Is that ocean warming in the Arctic Ocean melts the icepacks and kicks off Ice Ages
and a significant reduction in sea level causes them to end

The Cycle Begins

It is the melting of ice in the Arctic Ocean that begins the cycle that ends in the world dropping into the next Ice Age. Why does it do this? The short answer is that the warming Arctic breaks the Jet Stream and a lack of a strong Jet Stream allows air masses to stagnate over the continents and get colder and colder.

This is exactly what we have been seeing for the last couple of years, a weakening Jet Stream that is causing air masses to sit in their source regions far longer than we have seen in previous decades. When these air masses finally move out of these regions they are colder and more intense than we have seen before. They have even taken to naming winter storms. Don't these people have anything else to do? Like forecast the weather? Oh yeah it's the computers that do all the work, the forecasters just talk. A hell of a job doing it I must add, talking I mean.

The Jet Stream is one of the primary reasons we see changes in weather, differences in speed in the Jet Stream create upper level disturbances that propagate around the globe. These disturbances are imbedded in the Jet Stream and as they travel around the world they often cause surface air masses to become unstable and begin moving away from their source regions. These air masses tend to move east and south ward. It is these moving air masses that create much of what we consider to be weather. A healthy Jet Stream is critical in preventing air masses from staying in one location too long and the air masses becoming either too cold or too warm.

When the Arctic Ocean becomes mostly free of ice, the Jet Stream splits into at least two or more distinct jets that circle around intense glacier cold air masses located over Europe and Asia, and North America. These intensely cold air masses act as huge blocks and prevent the weak Jet Stream from circumnavigating the globe, resulting in the cold air masses becoming locked in place, ice then builds up under these extremely cold air masses and glaciers form.

This is why the ice does not melt.

Once glacier these cold air masses form they act as reservoirs of very cold surface air (anywhere from -10 to -70 Fahrenheit) and ice builds rapidly on their periphery, eventually forming huge glaciers wherever these cold air mass face an ocean (or any large water area). These glaciers build in such a manner that they gradually act more and more like a bowl, holding in the cold air of the air mass that was very cold to begin with and then gets colder and colder.

How long is this going to take? Once the glacier air masses get locked in place, the process begins and builds for tens of thousands of years. Climate over these regions rarely see a summer in the first couple of hundred years and none at all when the process gets going full blast. It will get cold and stay cold, during summers it will rarely warm enough in the southern regions to break the surface inversions.

Graphic 8 shows how air flows northward from these intense High Pressure systems over North America, Europe and Asia. These are surface winds, at high levels we a reverse flow of winds. In addition we often see low level jets that can reach speeds of 35-70 miles an hours. These are winds that ride over the surface layer of cold air. This is a main transporter of moisture to the building glaciers.

Glaciers form around these cold air masses on the sides facing oceans, mountains chains act as natural barriers that hold these extremely cold air masses in place, which create their own weather patterns and clouds.

A glacier is shielded by a thick inversion layer of stratus and ice fog that keeps the surface layer of ice and immensely cold surface air from the warming rays of the sun. An inversion is a condition where the temperature gets warmer as one rises in height, with the warmest layer being at least 1,000 feet up and the temperature dropping off rapidly above and below. To break out of an inversion, the cold surface layer has to warm the difference between the lowest surface temperature and the top of the inversion layer. During the coldest period of the last Ice Age, it takes the surface layer to warm by 50 to 100 degrees Fahrenheit (or more) for the inversion to break and for the ice to be open to the sun (this rarely happened).

Like a teeter totter Nature seeks balance,

a change in one area

often results in a corresponding and opposite change in another place,

but in many respects this is completely random

- For example -

We do not know if a fart from a water buffalo in India helps to produce rain in Iowa, but it certainly might.

CHAPTER 4

THE ON/OFF SWITCH

The oldest contiguous ice we have found so far in the glaciers of Greenland was created about 400,000 years ago. Why is this? The oldest ice we find in Antarctica is twice that age at 800,000 years. The age of the ice in Greenland corresponds to the start of the last four Ice Age cycles, so the question rose in my mind. "What happened 400,000 years ago to set off the repeating cycle of Ice Ages in the northern hemisphere"?

Prior to 400,000 years ago Greenland was not covered in glaciers. The science of tectonics tells us that Greenland is currently moving away from Europe at about an inch a year. Only 400,000 years ago did the opening get large enough to allow enough warm water into the Arctic to thaw the ice.

During the 4 ice ages the glaciers on Greenland and Europe were very heavy and pushed down the continental plates in these areas so much that this created an upward force in the straits between Europe and Greenland, resulting in the ocean floor rising by hundreds of feet and less water being able to transit these straits. On the continents where glaciers resided, the earth is still rebounding (with occasional earthquakes) from not having all that weight on them.

Ocean cores taken in the Arctic Ocean show periods of open water in the Arctic Ocean during the last Ice Age, caused by strong upwelling currents around the periphery of the Arctic Ocean. Meteorologically this upwelling resulted from near storm force surface winds flowing northward from the continents, as these winds move over water they blow the surface water towards the pole and this brings warmer water up from the depths. The spin of the earth causes these winds to veer towards the right but the results are the same.

As moisture is evaporated into the air above the Arctic Ocean, two things happen.

1. Cold air coming off the continents is very dense and colder than -40 degrees Fahrenheit, this are warms and this causes the air to become unstable and lighter than surrounding air.
   It rises and results in lower surface pressures.
2. As the winds evaporate moisture from the ocean surface: the water becomes much colder and salinity increases. Today this water will either freeze or sink, but due to surface mixing from the high winds churning the sea, the water does not freeze instead it gets pushed towards the pole and eventually sinks and becomes bottom water in the Arctic and Atlantic Oceans .

Water of one temperature or density does not really mix well with water of other temperatures and densities, this prevents the dense sea water from sinking, but at some point large areas of this cold saline (salty) water becomes far heavier than the water under it and globs of it sinks; falling until it finds equilibrium at the bottom of the Arctic Ocean. This cold dense water becomes part of the Arctic Ocean Deep Water Mass, shown in Table 1 as being colder and more salty than the incoming Atlantic in-flow.

Type	Temperature Fahrenheit	Salinity Parts per Thousand	Depth in Feet
Arctic Surface Water	28.6 to 30.2	28.0 to 34.0	Surface to 600
Atlantic Water (middle)	35.6 to 43.0	34.8 to 35.1	600 to 3000
Arctic Deep Water	30.6 to 35.6	34.9 to 35.0	3000 to bottom

Table 1. Water Mass Types and Characteristics of the Arctic Ocean

As gale to storm force winds blows northward over the Arctic coastline, upwelling brings warm Atlantic (middle) water (shown in Table 1) up. This warm water is cooled by these -40 to -70 Fahrenheit winds and only extreme mixing in the upper levels of the ocean keeps the water from freezing, the upper level of the ocean continue to become colder and denser, eventually sinking to the bottom of the Arctic becoming Arctic and Atlantic Deep Water.

This is an extremely efficient water cooling system and the resulting dense cold water eventually flows out of the Arctic by way of the western portion of the Greenland straits, between Iceland and Greenland. Even today this flow of water is one that we could harness for energy.

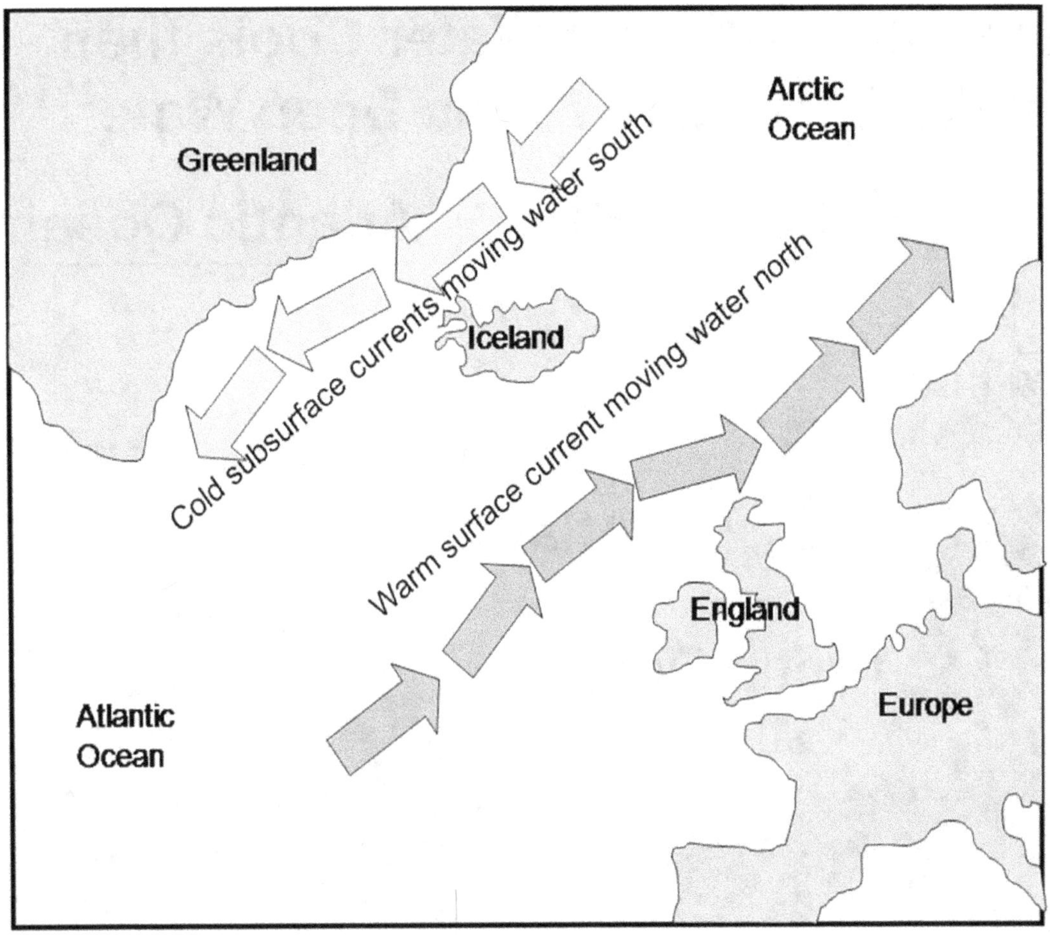

Graphic 5. Greenland / Iceland Straits, with Surface Warm Current and Cold subsurface currents

Taking a look at Graphic 5, every day a subsurface current (an offshoot of the Gulf Stream) pumps millions of square meters of warm water into the Arctic, this water becomes warm middle level Arctic water.

Nature loves equilibrium and to maintain a balance of this inward flow, millions of square meters of cold Arctic Deep Water flows out of the Arctic in dense subsurface currents, eventually becoming Bottom Water in the Atlantic basin. This outflow of dense cold Deep Water continues because the Arctic Deep Water is continually replaced by cold surface water dropping from the sea surface to the bottom of the Arctic. Even today this current is strong, but during the Ice Age it was much stronger, becoming a continuous freight train pulling cold water from the Arctic Ocean.

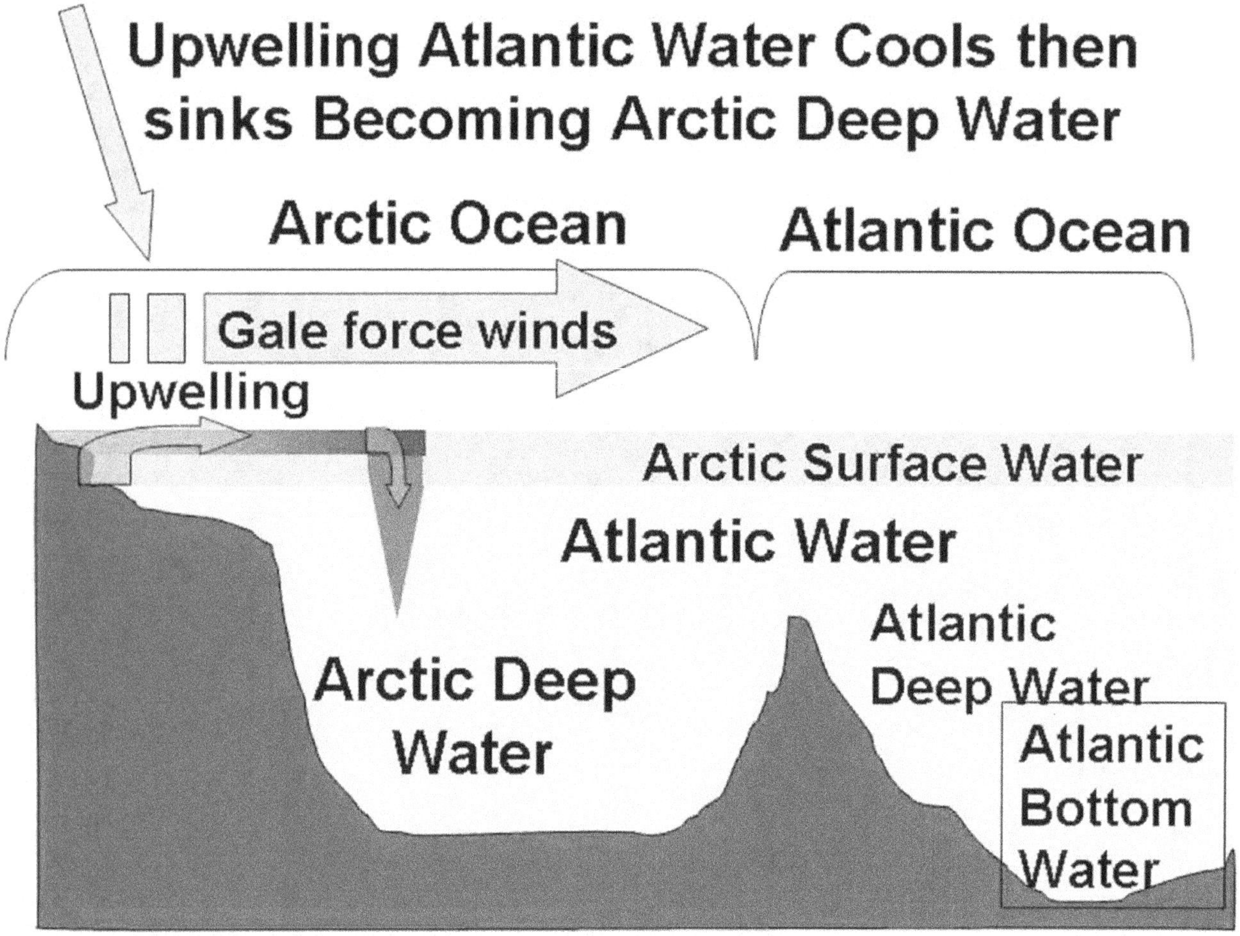

Graphic 6. How Upwelling Atlantic Water becomes Arctic Deep Water

During an Ice Age, continuous subzero (-40°F to -70°F) storm force winds blow directly offshore towards the Polar Low as depicted in Graphic 6, resulting in surface water getting blown towards the center of the Arctic and upwelling near the coastline. The spin of the earth induces a Coriolis effect, resulting in these winds turning towards the right, feeding into the low pressure center at the pole.

These violent winds cause huge waves and much vertical mixing in the surface layers of water, this prevents the water from getting either too cold or salty (dense), cooling the entire surface layer. This entire surface layer of cooling water flows north ward in a current that gets stronger and deeper as it gets further north, winds also increase the further north it gets. Portions of this cold water layer become denser than warm underlying water and globs of it sink through this warm water down to the bottom and become Arctic Bottom Water.

Some water sinks on the way to the pole, but the majority of this now dense and cold water eventually reaches the north pole and sinks in a vast whirl pool that carries this water down and towards the western Greenland Icelandic straits in a strong underwater current where the dense and cold water current exits the Arctic in a rapidly flowing subsurface current a mile deep and fifty wide.

This same effect happens all around the Arctic Ocean resulting in a vast amount of warm water coming into the Arctic being cooled into Arctic Deep Water and getting forced out of the Arctic by the whirlpool and underwater current. Debris from this must cover miles of ocean floor for thousands of feet.

These areas of upwelling bring nutrients up from the deep and support a vast array of plankton and larger animal life.

Upwelling

A sustained wind blowing directly offshore blows surface water seaward this water has to be replaced from somewhere and it causes upwelling water coming up from the depths to replace water that is getting pushed out to sea

In the Arctic Ocean, upwelling brings warmer Atlantic Water up from below.

The Tipping Point Switch

The Greenland / Iceland / English straits play a key role of regulating the temperature of the Arctic Ocean. Every day many millions of square feet of warm water flows northward east of Iceland, and a corresponding amount of cold water flows southward into the Atlantic west of Iceland. During the Ice Age the amount of water transiting these straits is many times today's volume.

When upwelling begins the first winter the Arctic Ocean becomes mostly free of ice, the warm middle layer of water in the Arctic Ocean is upwelled, the North Pole warms up and we are in for a very long cold winter.

We may be able to stop an ice age

Technology may be able to save us; we have an ON/OFF switch. The end of the last Ice Age (and four previous Ice Ages) occurred when either of flow of warm water into the Arctic or the flow of cold water out of the Arctic was interrupted. If we can interrupt the flow of water either into or out of the Arctic Ocean we may be able to avoid another Ice Age altogether or cause it to be short and brief. This is something we could do today but requires massive amounts of material and (the most difficult) cooperation of many nations.

Stopping or slowing the amount of water flowing both in or out of the Greenland / Iceland / England straits requires filling up the straits with hundreds of feet of gravel and rock, enough to stop or slow the flow of deep currents carrying water out of the Arctic. We also may be able to build floating surface or subsurface baffles that prevents so much water from entering or exiting the Arctic.

This might be the time to build floating hydroelectric power generators that harness the warm water flowing into Arctic or submerged subsurface power generators that harness the cold dense currents flowing out of the Arctic.

Air Pressure

Birds (and helicopters) are affected by the surface air pressure -
they can fly higher and easier when the pressure is high.

If birds are flying high in the sky, weather is usually good.

If smoke from smoke stacks rise easily,
the atmosphere is unstable, rain and storms may occur.

If smoke lays out flat or even sinks,
conditions are stable and weather will generally remain the same

CHAPTER 5

HOW WEATHER HAS CHANGED MAN

If humans have no other redeeming features, we adapt quickly to changing conditions. We live all over the world and subjected to extreme variations of heat, cold and elevation. For instance in Mongolia the yearly temperatures range from -65 degrees Fahrenheit in the winter to over 120 degrees Fahrenheit in the summer we easily survive and thrive in this hostile environment, modifying our habits and methods of scraping a living off the land.

Mongolia is as hostile an environment as it gets. It is not customary to knock when a stranger comes to a desert home (called a ger), you are welcome to enter without knocking and be offered food, drink, and shelter. One is supposed to call out and ask them to hold the dog. So in an extremely hostile environment we reacted by offering greater hospitality to strangers. Their culture is complex, for instance the very colors that one wears indicate social status. It was just a few hundred years ago that these friendly Mongols also conquered most all of Asia and part of Europe, they only quit heading west because the Khan died. The largest empire that ever existed.

During the last period of maximum of global warming, man spread over large portions of the northern continents; hunters and food gatherers saw this period as a dangerous time of plenty. An era that has been completely lost to us, few traces of these wonderers has been found. They may not have used complex tools, or weapons; but like us they certainly had the ability to adapt to their changing environment.

During this period so long ago, humanity developed language and social skills that allowed us to pass knowledge on to our children and others. Except for the lack of fast food and no beer, the earth was a paradise (as long as you were not eaten by a predator). Then the ice covering the Arctic Ocean melted, the upper level Jet Streams became disrupted, and extremely cold surface air began to pool again over Europe, Eurasia, and Canada, and ice again came to the northern continents.

The Ice Record is not precise as it covers 400,000 years. But one thing is clear, it does appear as if one summer it was really warm and the Ice Age began the next winter with little to no transition time. The previous four ice age maximums peaked and dropped right off. But this time man has thrown his wrench into the works and we have already modified weather for 10,000 years. There is really no telling what affect this is going to have on climate.

When the last Tipping Point was reached the world was once again plunged into ice. This made survival difficult and many were literally stranded in a world of ice. The transition from green to white was traumatic, no different from being in a severe ice storm today; the only difference was that the ice fell for weeks, months, and years (Tens of thousands of years…) and the ice simply did not melt. This made every factor of life dangerous like finding water to drink or simply walking was dangerous.

When ice invades the life of a hunter or food gatherer, the quick solution is go elsewhere; but this was not an easy decision in the world of 128,000 years ago. For Ice Age man around every bend and over every hill lurked new dangers and new terrors. Man was not the top predator of the day and the widespread use of fire most likely occurred 40,000-90,000 years later; at night for the time being we were fair game to carnivores that were both very hungry and also experiencing severe ice stress.

This was a world of complete unknowns, where escape from the ice meant every step was a step away from security and into a place where one had to relearn all the dangers that might be lurking behind every bush or tree. A world where ones homeland had a host of dangers, but these were known dangers. As soon as one leaves familiar territory the nights become increasingly terrifying; with unknown threats and new sounds (it is very important to know the sounds of danger in a world without fire). At night one has increasing trouble differencing between extremely dangerous threats and non-threats, terror prevails. No wonder the sound of fingernails on a chalkboard makes us uneasy, it is ingrained in us. So, just as today's man is locked in place by obligations, Ice Age man was locked in place by both familiarity of the old and fear of the unknown. It took fear of the ice, thirst, hunger, and the very sharpness of the cold to drive him into the unknown.

For modern man who is anchored in one spot by family and responsibilities; ice is just as difficult to cope with. Even a thin layer of ice in a modern city is enough to bring it to a virtual standstill. Imagine a foot of ice on Chicago that does not melt and only gets thicker; it devastates the city and every standing structure eventually fails under the weight of building ice. We build our skyscrapers one floor at a time and as we learned from the World Trade Center it only takes one floor to fail and even the tallest and mightiest skyscraper will fall. Almost nothing we make is impervious to ice; it seems as if ice is the perfect tool for nature to erase all marks of man from the face of the globe.

Man has the special ability to learn, we can pass experiences down to our friends and descendants; we learn to cope with things as a group and to fight as one. It is why the human race has been able to cope with changes that wiped other less adaptable species off the face of the planet. It is why we survived the ages as we are definitely hard to kill. But our dwelling and other creations have a much more difficult time coping with the hardships of time and a changing environment; only gold, a couple other metals, and pottery are tough enough to last tens of thousands of years. Few items predate the end of the last Ice Age and these are usually weapons such as stone spears or arrow heads, but remarkably some exquisite gold jewelry has been found in Germany that is 30,000 years old, someone made this.

17,000 years ago the planet was a different place, glaciers still covered a good portion of the northern continents, yet the ice was beginning to recede. The Jet Stream had finally returned to the middle latitudes of the northern hemisphere and the upper level westerly winds brought moisture and warmth into the cold air masses at the center of the glaciers, gradually removing the chill from the heart of the glaciers. This warmth eventually melted the glaciers from the inside out. It must have been an awesome sight watching mountains of glacier ice melt into vast raging rivers, with vast floods beginning and ending almost overnight, depending on the size of the melt water pools as trapped bodies of melt water gradually find their way out of the ice through cracks and channels in the glaciers, these channels rapidly enlarge and huge floods occur. One day a huge lake of melt water may sit on the ice and the next day be gone. We see this same effect today in Greenland as the glaciers melt.

Life in the shadow of a glacier

For the next 110,000 years some of our ancestors lived in the shadows of these mountains of ice, life was difficult and everyday activities dangerous, if the wind simply shifted direction the extremely cold air from the glaciers could drop temperatures from above freezing to deadly cold in an instant. This is like opening the freezer door and feeling the cold air hit your feet. These are Katabatic winds (a Greek word) In the case of glacier winds, the air from a glacier is moving at category 5 hurricane speed and can drop temperatures (in seconds) by a hundred degrees or more.

The area between the glacier and its water sources to the south was very fertile and warm as moist winds routinely blew from the sea towards the glaciers. Occasionally these winds shifted 180 degrees and what as a beautiful warm day ended up with temperature drops of 50 to 100 degrees Fahrenheit. What had been a warm and fertile grazing area with lots of herbivores calming grazing turns into a deadly wasteland as it instantly became deadly cold, any unprotected animal or person was flash frozen by savage winds of frigid death. Humans adapted to these harsh conditions and thrived by scavenging on the frozen victims, competing with many carnivores for the choicest picking. Interesting enough, remains of whole mammoths have been found in the thawing thermafrost of Siberia that still had grass undigested in their stomachs, appearing to of been flash frozen.

Some animals such as the Sabretooth tiger developed specialized tools to eat the flesh of frozen animals, their large saber fangs were used to scrap parallel vertical channels in the meat of frozen carrion, while their front teeth (like molars) allowed them to scrape a layer off at a time from between the groves. Instead of being top level predators, these large Sabretooth cats survived on frozen victims of the ice. Not that they couldn't kill, but why run and risk injury when you have a banquet frozen in place? This is part of the reason they died when the ice melted, they lost their food source.

Fire has taken its toll, Eruptions

The minor eruption of Yellowstone Caldera 116,000 years ago took away the sun for many years and threw the globe into a volcanic winter lasting 8,000 years; a period when global temperatures plummeted.

Much of humanity migrated southward and was concentrated along the coastlines of both the Mediterranean Sea and the Indian Ocean where we flourished. Then 30,000 years later Toba Caldera (74,000 years ago) erupted causing huge earthquakes and monster size Tsunami in the Indian Ocean, this big eruption took away the sun for months.

In both eruptions of Yellowstone and Toba global temperatures dropped many degrees and pollutants almost completely destroyed most of the world's protective layer of ozone, allowing deadly ultraviolet rays of the sun (that the ozone layer normally protects us from) to kill any plants that survived the months of darkness. After a long period of darkness and hunger, only the most fit and able of survivors ventured forth to greet the rising sun, a joyous experience to view after months of darkness, but this celebration was short lived as they found most vegetation dead or dying from lack of sun and they were gradually blinded by deadly ultraviolet rays of the sun. These same rays made short work of the remainder of the world's plants. Man, beast, and plant succumbed, only those living in very cloudy climates such as near glaciers and on the windward side of huge mountain chains avoided this fate. DNA analysis has revealed that mankind was nearly wiped out, only 1 in 10,000 mating female's survived this period in our history.

The Ice record shows the world dropping into an 8,000 year volcanic winter after Yellowstone erupted. Any drop that happened after Toba erupted must have been of very short duration as the ice record shows only a small drop in this time frame, it almost seems as if no eruption occurred in this period.

They must have been very different types of eruptions, I theorize:

1. Toba went off like a grenade, all at once, very explosive, and caused catastrophic climate changes that lasted for a few years. This was like popping the cap off a shaken soda bottle, all at once. This is reflected by the Ice Record.

2. Yellowstone on the other hand erupted for years, delivering much more steam, pollutants, and magma into the stratosphere for a much longer period of time. Climate contributed to this eruption as Yellowstone is at the end of a long valley that brings moisture from the Pacific Ocean, the area directly over the caldera quickly built up several miles of ice at the start of the last Ice Age. This ice built up over 15,000 years and was a lot of additional stress on the caldron of magma far below. Not enough stress to cause the entire caldera to explode but instead causing a series of non-catastrophic eruptions that allowed magma to escape the caldron in the chamber below in a series of Plinian eruptions lasting hundreds of generations.

"Laacher See" Caldera Volcano

Between 11,500 to 12,286 years ago; right at the height of the melting of the glacier ice, the world abruptly stopped warming and plunged back into a mini ice age for 700 years. This period is called the Younger Dryas and shows up on the Ice Record as seen on Graphic 2. During this period the "Laacher See" (a small caldera volcano in Germany) had a number of Plinian eruptions, these are very dramatic events that send magma columns over 25 miles into the sky.

The Laacher See is located in Germany and is geographically positioned in such a manner that a Plinian eruption has a huge effect on the Albedo of the entire northern hemisphere. These eruptions blasted quite a bit of dust and gasses into the stratosphere that took many years to dissipate, this is reflected by the amount of dust in the Ice Record for this period. This dust and volcanic gases caused a good portion of the incoming sunlight to be reflected back into space and allowed the melting glaciers to refreeze. This had little effect on already frozen Arctic ice coverage (the Arctic iced back over around

17,000 years ago), thus the Jet Stream was in great health. Then the warming started again as soon as the dust and gases dissipated. This is shown by the Ice Record.

Highly paid and respected scientist at NOAA (likely with real bad hair AND mismatched socks) blame the fresh water from melting glaciers for this 700 years of cooling, explaining the influx of fresh water into the North Atlantic from the melting great ice sheets caused a reduction in the ocean's thermohaline circulation and this in turn caused this short period of global cooling. They make no mention of the eruption of Laacher See. I think they are out of their minds and putting the cart in front of the horse as the warming climate melted the ice which interrupted the oceans thermohaline circulation; this recovered soon after. In addition we had three previous warming periods in the last 400,000 years and the influx of melt water did not have this warming effect . This had little effect on global cooling and no effect of the Jet Stream. It's changes in climate that come first and affects the ocean. Not the other way around.

The Beach

On a day when waves are breaking, it is possible for freak waves to come ashore,
waves that are much larger than other waves.
They do not often occur – but do happen
During a hurricane,
the beach is not the place to be, especially a category 4 or 5 storm.
Freak waves are much more prevalent in such a hurricane

If you go to the beach in a hurricane, never turn your back to the sea,
and don't get anywhere near the break zone.
If you do, a freak wave can easily take you away.

In fact,
don't ever turn your back on the sea.

CHAPTER 6

ICE

Modern man looks at ice as a necessity of life and absolutely nothing is more important on a hot summer day to the modern day descendent of Neanderthal man than an ice chest full of ice and beer and going to see a sporting event, this is a rite of passage. Why does ice do such an awesome job of cooling our beer and other assorted drinks? It is after all simply water whose molecules are locked tightly together with hydrogen bonds. The magic occurs when warming breaks these bonds, it takes a lot of heat energy to free these bonds and this energy is absorbed from the surrounding water.

Throw warm beer into an ice chest full of ice and the warmth of the beer radiates out into the surrounding water and ice. The ice absorbs this heat energy and hydrogen bonds are gradually broken. It takes 80 times the energy to melt 1 gram of ice than it does to warm 1 gram of water by 1 degree Celsius; changes of state are every expensive in terms of the amount of energy required. In the case of our ice chest as heat from the beer radiates into surrounding ice, the heat loosens the bonds between the ice molecules and a change of state occurs as ice melts into water; hydrogen bonds are slowly weakened and broken and this process continues and absorbs copious amount of heat energy from the surrounding water. This is the process that cools beer.

Cooling of Ice Covered Sea by Cold Winds

Ice is a GREAT insulator between the ocean and the sky. When cold dry air blows over a body of water that is covered by ice; the ice makes it extremely difficult for the air to absorb moisture from the ocean below. The result is that air over an ice covered body of water is very dry, like a desert. By comparison, a sustained strong cold dry wind blowing directly over a body of open water absorbs moisture from the water below like crazy; this is near the perfect evaporation scenario and the air basically vacuums moisture off the ocean surface.

Once dry air absorbs moisture it becomes far lighter than the surrounding air, becoming unstable and rising, being replaced by more dry cold air that continues to absorb water from the sea. This process gradually builds a thicker and thicker maritime layer right next to the sea surface. First patchy fog forms, then as the air absorbs more moisture it builds into a thick layer of fog, stratus, and stratocumulus (usually a combination of all three); creating a thick maritime layer.

Two very unique physical factors of ice:

1. It floats because it has much less density than normal water.

2. It retains its form (ice crystals) very well and takes a lot of energy to melt.

In an enclosed body of water such as the Arctic Ocean the mere existence of a layer of ice on the surface has a profoundly stabilizing effect on the atmosphere above the ice. The air over the Arctic Ice is much like the air over a cold dry desert (Mongolia in the winter): dry, stable, and intensely cold.

Ice by definition is at or below 32 degrees Fahrenheit (28.4 degrees Fahrenheit for Sea Water). In the ice covered Arctic Ocean this creates a stable layer of air near the surface of the sea ice. Ice is the great moderator as colder air moving over ice will gradually make the ice colder, while warmer air moving over ice will rapidly cool the air to the temperature of the ice. In addition, ice does not evaporate very well, it takes about six times as much energy to cause water to change (sublimate) from ice to vapor as it does for water to just evaporate off the ocean surface. This results in a layer of stable and extremely dry and cold air sitting on the ice pack.

Magical Ice

Water and ice are identical molecules, but act very different in how they are affected by heat or cooling. Changes of state are expensive in energy (ice to liquid or liquid to vapor).

Liquid water: it takes 1 calorie (different than food calories) to raise 1 gram of water 1 degree; conversely 1 calorie of energy is released when 1 gram of water lowers by 1 degree (this relationship is shown by Graphic 7).

Ice: To simply convert 1 gram of ice into water (at the same temperature) takes 80 calories

Ice is a reverse energy sink. The hydrogen bonds between the ice molecules are strong and a reflection of the cold that created them. Unlike water that absorbs and sheds heat quickly, ice absorbs 80 times the amount of heat energy before it melts. Now imagine a chunk of ice 3 miles thick and minus 70 degrees Fahrenheit that covers millions of square miles, it has a huge capacity of absorbing heat before it even thinks about melting. Just melting the surface was a great feat.

This capacity of ice to absorb huge quantities of heat before melting was a big reason why the Ice Age lasted as long as it did; as it simply takes a lot of heating to break the icy grip that glaciers had on our continents. More energy than sunshine alone can provide, it takes a continuous warm wind bringing moisture and heat from the ocean.

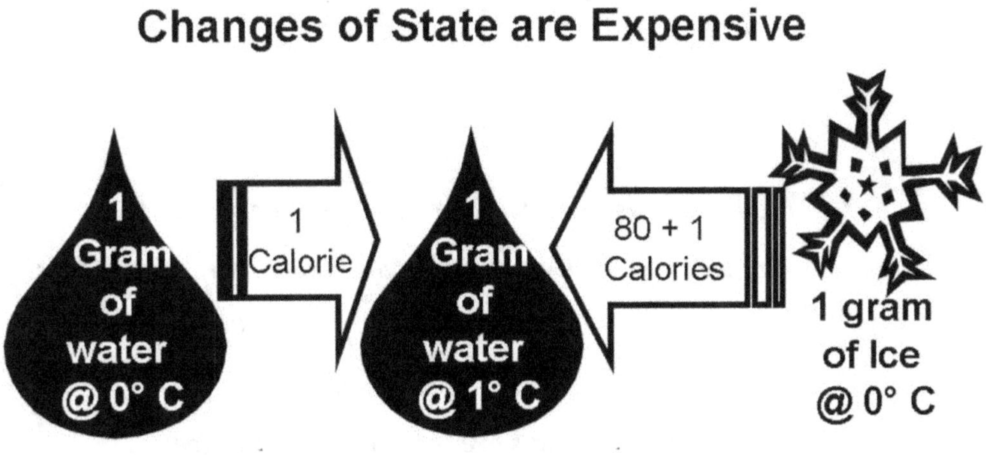

Graphic 7. Ice holds energy hostage

Looking at Graphic 7, it is seen that it takes 80 calories to simply melt 1 gram of ice and 1 more calorie to raise it 1 degree.

On a global scale, if on a given day 1 trillion grams of ice cover melts, the amount of heat absorbed to do this is 80 trillion calories. After the ice melts and the ground is now ice free, the sun heats the ground instead of melting ice. The ground heats and transfers this energy to the air above quite effectively. The same amount of heat energy (80 trillion calories) that on the previous day melted 1 trillion grams of ice, is now heating the ground, which absorbs most of the same 80 trillion calories of energy, the temperature of the ground is raised several degrees and radiates much of this energy to the air next to the ground. This contributes directly to global warming.

Simply put - As more ice melts.

The world gets much hotter, much quicker.

Ice

Fresh water is most dense at 40 degrees Fahrenheit,

below this temperature ice crystals begin forming and these decrease the density of water

Ice – thank god – floats

Or else the world's oceans would be solid ice

CHAPTER 7

AIR MASSES AND THE LIFE OF A GLACIER

After the last temperature maximum 128,000 years ago the Arctic Ocean became free of ice and the northern continents got very cold. This created a situation where warm air (low pressure) over the pole was surrounded by much colder air masses (high pressure) over the continents. The existence of a Low over the pole and intense areas of cold air over the continents caused the Jet Stream to completely disorganize, allowing the cold air masses over the continents to remain stationary and keep getting colder (a strong Jet Stream is why air masses move from one region to another). Ice was deposited and glaciers gradually developed under these stationary air masses.

When very cold air lays over the continent, the cold air acts much like oil on water – filling all low laying areas with a smooth upper surface that acts as a smooth highway for warmer air to flow over. This removes friction and often allows a Low Level Jet to develop that allows moisture from the Gulf of Mexico to flow northward to Canada, moving at 30 to 50 miles an hour – right over the underlying cold air and traveling many hundreds of miles in a 24 hours.

In today's winter, a Low Level Jet is often responsible for ice storms as warm moist air moves northward over underlying cold air - moisture precipitates out of the warm air into the cold air and freezes as it falls. This requires at least 1000 feet of cold air. Much the same as it did during Ice Ages, except during Ice Ages it does not melt.

The birth of a glacier

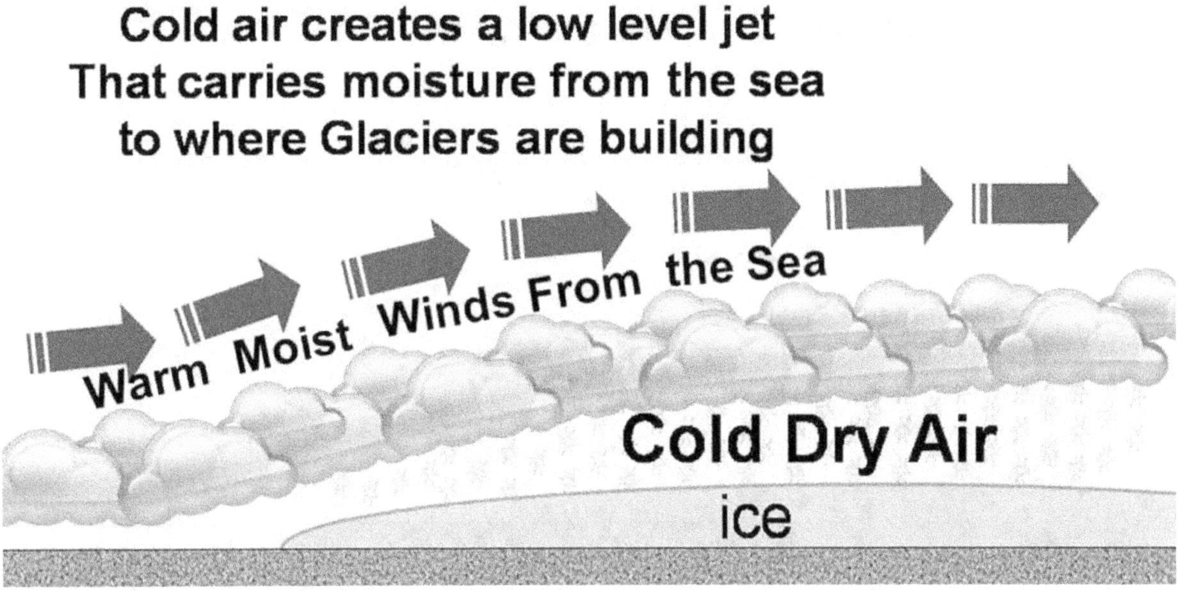

Graphic 8. Low Level Jet riding up and over an under laying cold air mass

Low Level Jet: intensely cold air covers the interior of the continent, filling all low laying terrain and creating a frictionless surface over which warmer air can ride over. Warm moist air rides over this cold air for hundreds of miles. The top of the dry cold air is a smooth and nearly frictionless surface on top of which warmer air can travel over with very little effort.

As winds from the south blow northward they bring moist warm air up and over the cold surface air mass, shown in Graphic 8 and the moisture falls as ice and snow. As an example of a Low Level Jet exists in Asia during winter: Mongolia is a cold desert that only receives 6 to 8 inches of precipitation a year, most during the month of February, this moisture comes from the South China Sea several thousand miles south. When the winds are just right this moisture rides up and over an underlying layer of cold air during the winter and brings welcome moisture to Mongolia.

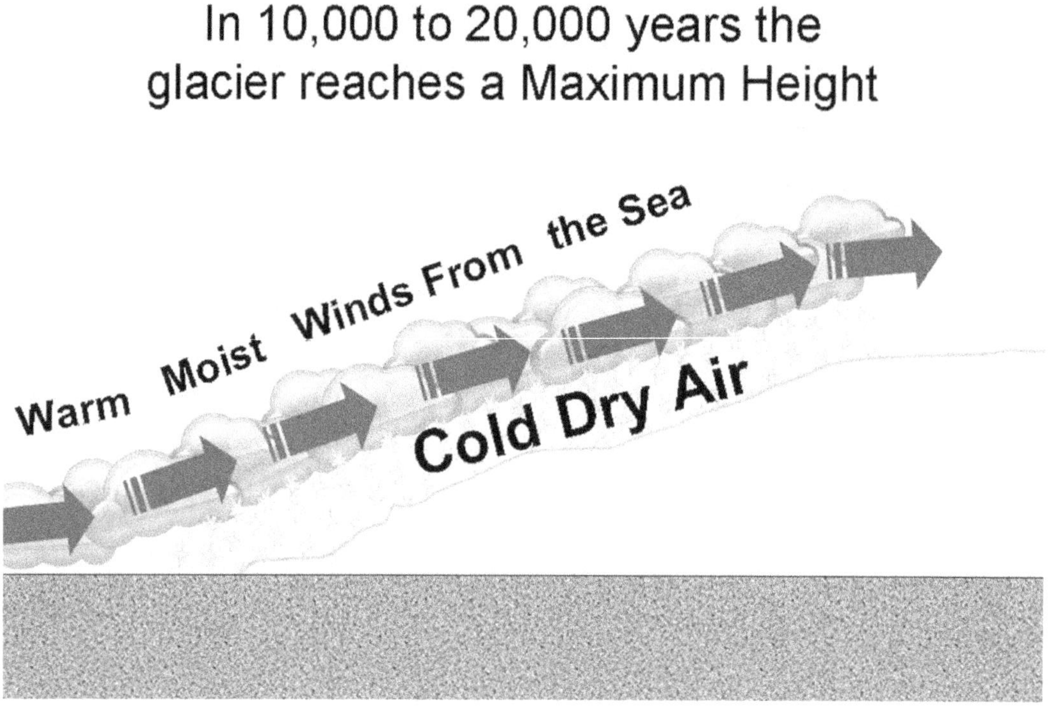

Graphic 9. Maximum height of a building glacier

Adiabatic cooling and warming

As moist winds flow up the side of a mountain they cool at the wet adiabatic lapse rate of 3.5 degrees Fahrenheit per 1,000 feet. If the mountain is tall enough the moist winds will cool to their dew point and moisture will begin to precipitate onto the mountain. If it is a tall mountain, at some point all the moisture falls out and the air becomes dry, this would be the tree line on a large mountain. This same effect acts on a glacier - as winds from the south continue to build ice for tens of thousands of years, these glaciers eventually reach a maximum height; this is the height where all moisture has precipitated out of the rising moist air, shown in Graphic 9. The glacier can grow no taller than this point.

It is the original temperature and the moisture content of the air that limits the height to which a glacier can grow. Both of these variables change as sea levels fall and moisture becomes less abundant, but the change is minimal unless the moisture source completely dries up. The maximum glacier thickness in Wisconsin was anywhere from 12,000 to 20,000 feet, while glaciers in the far north stood less tall at 8,000 to 12,000 feet; these maximum heights are totally due to the temperature and moisture content of the air source. They can grow no taller than the precipitation will fall.

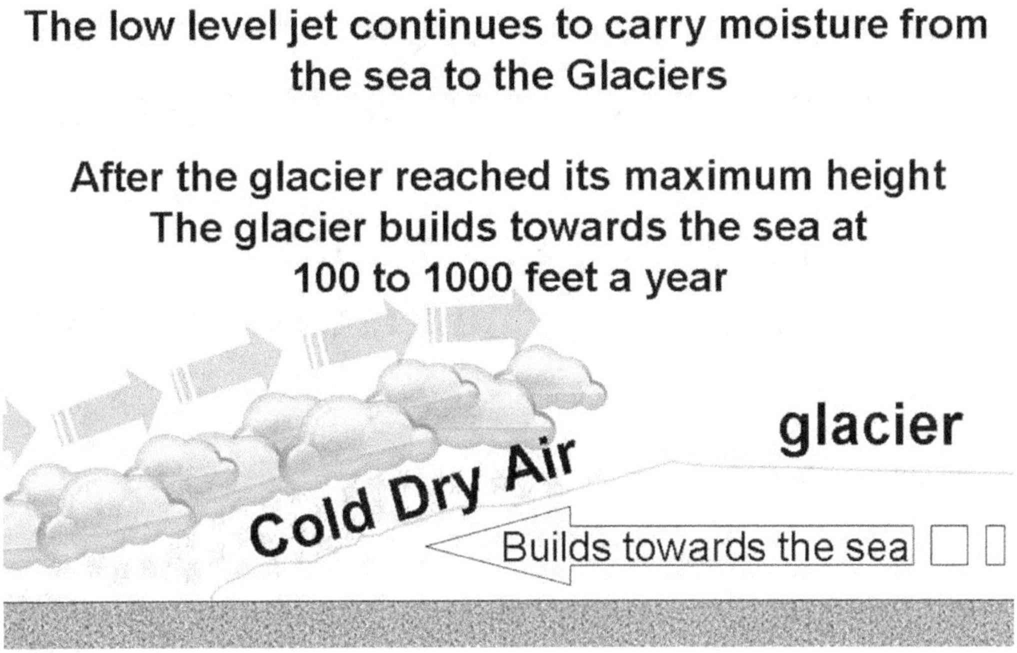

The low level jet continues to carry moisture from the sea to the Glaciers
After the glacier reached its maximum height The glacier builds towards the sea at 100 to 1000 feet a year
glacier
Cold Dry Air
Builds towards the sea

Graphic 10. The glacier starts building towards the moisture source

After a glacier reaches its maximum height, instead of growing upwards as in Graphic 8 and 9, the glacier begins growing rapidly towards the moisture source as shown in Graphic 10. All the moisture that used to fall over hundreds or thousands of square miles now falls entirely on the face of the developing glacier, the glacier now races towards its moisture source (well as fast as a glacier can slowly grow towards it anyway). First the glacier grows at high as it can, then the glacier begins to grow towards its moisture source. It should grow at least a 500-1,000 feet a year for a glacier feed by abundant moisture, Much like the Cascades of Washington, the windward maritime side gets abundant rain while the other side gets barely any.

The growing of the glacier face is dependent on the wind and amount of moisture in these winds. These winds vary, deep winter brings both the stronger southward flowing cold dry surface wind and an upper level wind that brings warm moist air northward. Glaciers facing the east and south side grow larger due to the warmth and moisture content of the moisture sources in the Gulf of Mexico and the

Atlantic. Glaciers in the far north are dependent on the Arctic Ocean for moisture and will not grow nearly as tall or as quick as their southern brothers.

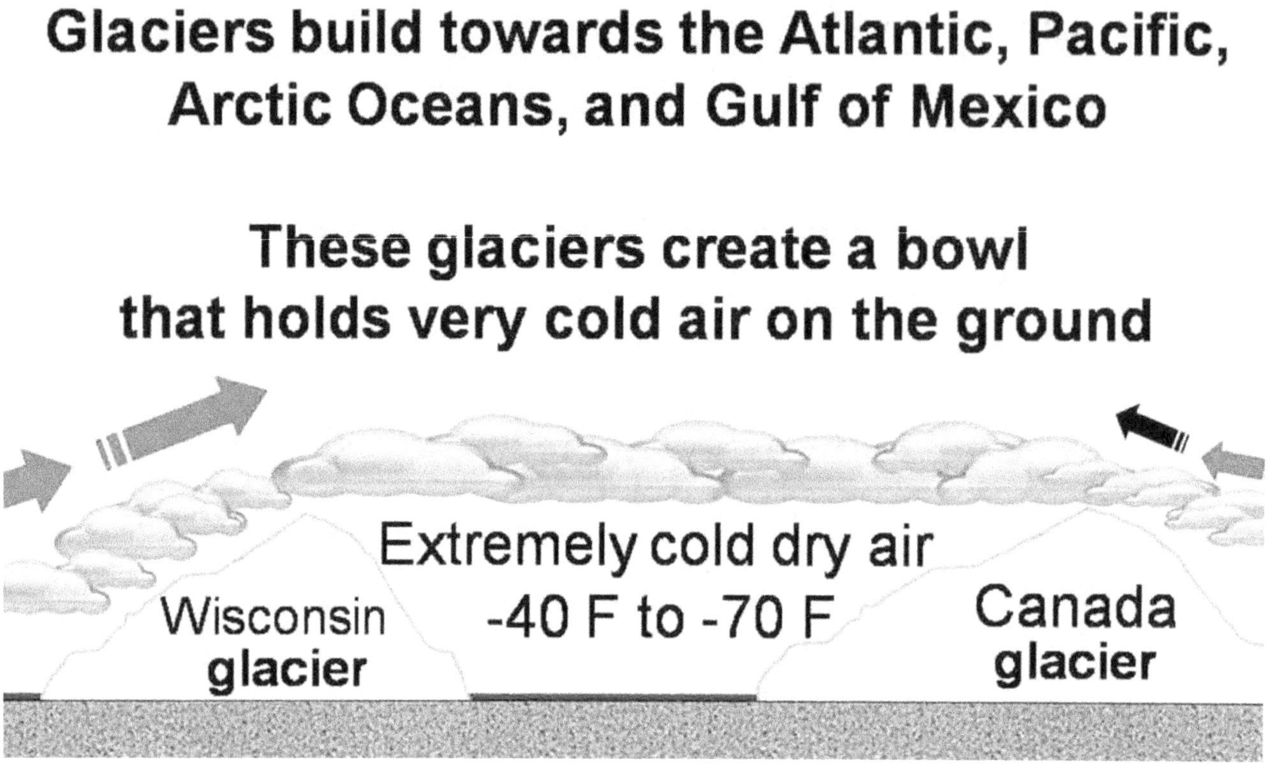
Glaciers build towards the Atlantic, Pacific, Arctic Oceans, and Gulf of Mexico
These glaciers create a bowl that holds very cold air on the ground
Extremely cold dry air
-40 F to -70 F
Wisconsin glacier
Canada glacier

Graphic 11. The cold air gathers behind the glaciers – creating a bowl

The glaciers building towards the seas in the south and others building towards the north creates a bowl of sorts, with very cold air collecting between glaciers to the north and south. This bowl collects surface air as shown in Graphic 11 by the big arrow between Wisconsin and Canada, mountain ranges such as the Rocky Mountains and the Appalachians help to create geographical barriers that hold this cold air.

Meteorologically these frigid air masses are very interesting, above these cold air masses lays a strong inversion. This inversion and surface fog and cloud layer acts to prevent sunlight from reaching the ground. This is a thick strong layer of stratus, stratocumulus, and surface fog that protects the underlying surface of the ice from the sun. In fact for this inversion to break and the cloud cover to dissipate, the surface temperatures have to warm as much as 50 to 100 degrees Fahrenheit. This huge inversion acts as both a protective layer and modifying layer, preventing diurnal (day and night) temperature drops or rises from reaching the ground.

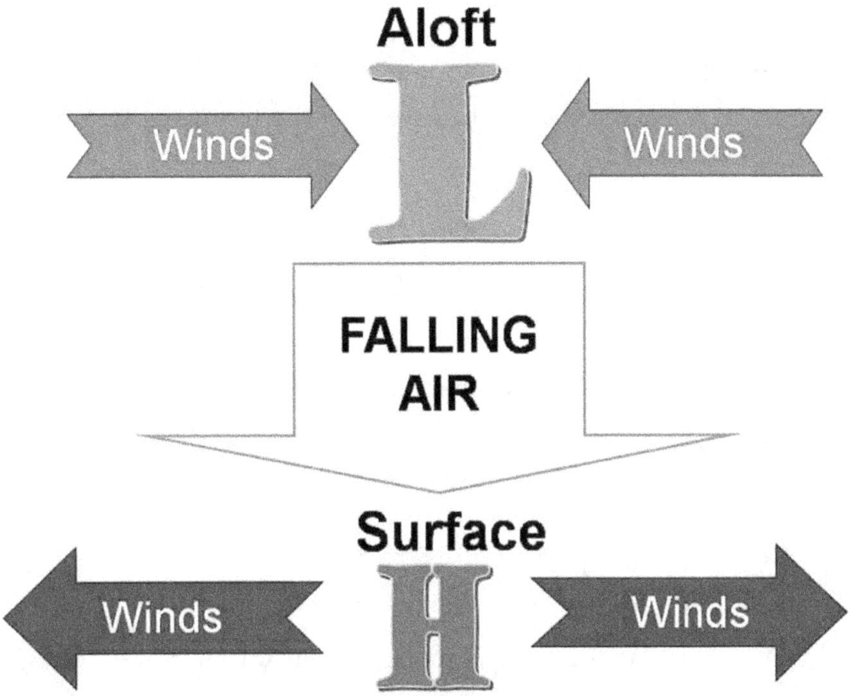

Graphic 12. Vertical conditions above the Glaciers, a cold core High with a Low aloft

A cold core high (and a cold core low) becomes colder towards the middle of the high, temperatures decrease and surface pressures increase. These cold highs rotate clockwise at the surface and often reverse at upper levels. This is shown in Graphic 12. In this manner Mother Nature achieves balance, many physical aspects of the atmosphere reverse from one level to the next:

- Where you have warm core High pressure at the surface - you have Low pressure aloft.
- Where you have cold core High pressure at the surface - you often have an area of low pressure aloft.
- Where air is divergent at the surface you have convergence aloft and the inverse is true. Air on the surface of a cold core High is very divergent, much like cold air falling out of a freezer when you open the door.
- Where temperatures are extremely cold at the surface, they stay the same aloft.

- An extremely strong and powerful cold core High pressure system at the surface results in low pressure aloft.

During the last Ice Age, the cold air on the surface of a glacier was far colder and of much higher surface pressure than any air mass that has been seen in the Northern Hemisphere in the last 16,000 years.

The bottom line: intensely cold air results in high density and very high surface pressure.

High altitude conditions over the glaciers have a strong upper level low pressure systems. It is these lows over these intense Highs that cause the Jet Stream to remain weak and disorganized.

The End of the Ice Age

At the end of the last Ice Age when the Arctic Ocean froze over, the Jet Stream finally returned to normal. This event occurred 17,000 years ago and kicked off warming and melting of the glaciers. This took several thousand years to breach the westerly walls of the glacier complexes, once that happened the glaciers started rapidly melting when warm westerly winds feed directly into the heart of the glacier complexes.

As winds brought warm moist air up and over the top of the glaciers, rain and warm winds gradually melted the faces of the glaciers. More importantly as this air flowed up and over the top of the glacier walls and down the far side of the glaciers, adiabatic compression of the falling air caused the temperature of the air falling on the lee side of the mountain of ice to increase in temperature. This warming effect is why dry and hot deserts are on the lee side of mountain chains. This heating caused the glaciers to melt from the inside out, creating huge amounts of melt water held back by the masses of ice still facing the oceans. When these lakes of melt water finally found cracks to escape, huge floods were unleashed that lasted weeks or months; such a series of torrents cut the spectacular Columbia River Gorge in Oregon. These cracks refreeze and the cycle happened over and over until the ice walls were fully breached.

Floods

Modern man has little recent memory of floods, but a distant relative of the biblical flood is the Christmas day 2004 Tsunami that killed over 200,000 people. This was a definitely a biblical class flood and one that will be remembered for generations. In many places this Tsunami was a series of waves over 30 feet tall. A fault line just like the one that generated the Christmas Day 2004 tsunami one lays off the coast of Washington and Oregon, it has been asleep for several hundred years and is due to strike. We don't know when. This is why new building codes in Seattle are so strict.

Our Esteemed Scientists have no idea how a Tsunami is born

They don't really know how Tsunami are generated. It's thought that the lifting of the crust during an earthquake lifts up the entire column of water to the surface and creates a bulge, then this bulge of water spreads out at 600 miles an hour and attacks distant shorelines with Tsunami waves that strike every 20 minutes or so. This in a body of water that is tens of thousands of feet deep. This is bull. Lifting will cause a bulge of course but a bulge is not going to kill people, it does not transfer enough energy to the water. Nor does it create a series of huge waves that hit every 15-20 minutes.

There has got to be something else and I'm convinced that much of the high intensity energy of the quake is transferred from the ocean floor into the water, these high intensity waves travel upward (the same reason that sound travels upward from the sea floor as energy waves abhor higher densities and travel nearly straight up to the surface). These high energy waves bounce off the sea surface and refract back down towards deep water where they becomes trapped in the deep sound channel. These waves of energy expand outwards in the deep sound channel and generate waves every 30 miles as they repeatedly bounce off the surface of the water and bend back up at the bottom of the deep sound channel.

In antisubmarine warfare, we utilize this same type of bounce to get extended ranges. It's called a convergence zone and occurs every 30 miles. This happens because sound waves (or any form of energy waves) traveling downward in the ocean are gradually bent back up towards the surface, due to energy waves tendency to bend away from higher densities. These waves converge 30 miles out and then repeat until the energy is spent.

Whale songs travel in the channel in the same way. In this manner multiple tsunami waves are generated by earthquake energy with a wavelength of 30 miles. The speed of a wave is related to its wavelength and a tsunami wave with a wavelength of 30 miles will travel 600 miles an hour. This is the exact speed that tsunami waves travel across the ocean. It's a rule of thumb. The first wave near the quake (nearest the epicenter is the last wave to hit land) is the strongest as it gains the most energy, with each successive wave generated getting weaker and weaker. This is exactly how they hit land, weakest first then bigger and bigger. The last one is the biggest. But I am not a Doctor and no one will even consider my idea. Getting used to this. Bastards.

It really irks me that so many died and we don't really know how these huge waves are generated. That is another book of mine.

I diverge again. Should have been a politician as I like to hear myself talk.

17,000 years ago, people lived along the coastline of a world where sea levels were 400 feet below what they are today. The Aegean Sea was much lower by at least 800 to 2000 feet. This is directly due to a combination of evaporation and the reduced inflow at the Straits of Gibraltar.

Islands out of nowhere

The lowering of the Aegean Sea was gradual as sea levels fell over 111,000 years. Islands first appeared as mere islets and then over time became bigger and bigger. These were free of the large predators of the time and soon became covered with vegetation. The first inhabitants of these islands found a land of warmth and safety with a bountiful harvest of fish and fowl. The weather was mild and the sun rarely broke through the thick layers of clouds, only during the summer did this happen. A land of plenty and the nursing place for a fledgling human race. It was our home for tens of thousands of years, we thrived and multiplied to hundreds of thousands.

It was an Eden, these islands in the Aegean Sea. Look at how far man has progressed in the last 500 years. Imagine we had a safe refuge from the dangers of the wild and developed unhindered over tens of thousands of years? If only a single man and a wife found Eden, at a population growth rate that doubles every 50 years they would reach a million in 2,000 years. This was a place that was safe for man for over 60,000 years. We multiplied. They had nothing else to do as drinking, the internet and Television had not been invented yet.

What wonders we could have achieved in 60,000 years. In many places around the world we see ancient wonders that have been in place for thousands of years, we don't know who built them. Many of these structures were made with stones carved with high precision and fitted together as if they were puzzles, some stones weighing hundreds of tons. So heavy that we cannot move them today. Who built these? It's easy to just disregard them and move on with our lives, but the fact remains that they exist.

Many believe (generally people with really bad hair) that aliens came to the earth and either made these puzzling artifacts we find all over the earth or helped us make them. This is idiotic, we are light years from the nearest star and the idea that intelligent beings from another star system traveled here to nurture our race is too much. Besides if they were here to nurture us, they didn't do a very good job. It is much more believable that in the far past we have developed higher technology than in use today but lost it in the mists of time and rising seas.

Our tools and homes are not made to last thousands of years, yet the ideas and beliefs that we hold true are a bit more enduring. The United States constitution is based on many ideas passed down from Greeks who lived thousands of years ago. Is it not possible that many of these ideas were passed down to the Greeks from a previous era? Perhaps the gods of old were just exceptional people with high technology, makes a lot more sense than someone able to throw lightning bolts (that would be cool).

17,000 years ago when the Jet Stream finally reasserted itself, it bought warmth to the heart of the great glaciers and they began melting from the inside out. Huge melt water lakes formed in the center of these glacier basins. These basins were like bowls holding trillions of gallons of ice cold water, for years these lakes got larger and larger until the water finally forced its way through cracks in the ice to escape. These lakes then rushed out in biblical size floods. Then the cracks refroze and this happened over and over as the glaciers melted. Over thousands of years the towering masses of ice became more and more porous and floods happened more frequently. Until the glacier walls were completely breached and a continuous flow began.

In the Atlantic and Pacific these floods created a sea level rise of a foot every ten years. In the first years of the rise in the Aegean Sea it was much faster, due to it being a smaller basin and being at a lower level (until the Aegean was nearly the same level as the rest of the world's oceans). At the start of the floods the Aegean rose two to five feet a week, slowly getting higher and higher. Perhaps 40 days was long enough for the main islands to go under and where we get the legend of Noah's flood.

The same islands that were a paradise for man for thousands of years became a deathtrap when the flood began. The sea rose higher and higher until all the islands became submerged. Imagine the horror as tens of thousands of families kept moving to higher and higher land all the while hoping the sea would quit rising, in many cases they had to decide who lives and dies. But in the end all perished except the few who were able to escape by boat or on debris left floating on the sea. This was the single greatest loss of life that mankind has experienced since the eruption of Toba 74,000 years ago, a civilization died. This was the place of Noah and the Ark. I view Noah taking in two of all the animals a crime, as this was definitely a choice between a boat load of people or animals.

I am sure we can find some ruins of this civilization. I just don't think anyone has looked in the right place, look in the Aegean. Look for ruins on the tops of sea mounts, look in deep water 200 to 2,000 feet down. If they built with wood, nothing is left. But I expect they used a lot of stone in the same manner as the Greeks (who had to learn it somewhere). We may even find Atlantis.

This same buildup of melt water and resulting floods was how glaciers north of the Black Sea failed, resulting in huge floods into the Black Sea (where the ruins of underwater villages exist) and after this basin was full, flooding onward into the Mediterranean Sea. The Straits of Dardanelles had two floods, one when the first floods came from the north from the glacier melt water and spilling into the Mediterranean and filling the Aegean; then another flood much later when the Aegean became level with the Atlantic, this flood went the other direction. Almost every ancient people around the globe seems to have legends of floods; this is reasonable as glaciers existed around the globe and this same effect caused flooding in many areas. In Oregon the Columbia River Gorge in Oregon was cut by a series of such floods, it is 10 miles wide and 2 thousand feet deep.

Why Ice Ages occur

It was lack of very cold air in the Arctic Ocean that created conditions over the continents suitable for the formation of a type of air mass not seen in the last 17,000 years; this is the permanent continental glacier air mass having the following properties:

1. Intensely cold, anywhere from -40 F to -100 F. coldest where the pressure was highest and upper level subsidence is greatest.
2. Extremely dry, frost forms in your freezer because cold air does not hold moisture well and the colder it is the harder it is to hold any moisture.
3. Extremely dense; cold dry air is very dense, more dense than any seen in nature today.

Inversion

When very cold air has warmer air above, it is called an inversion because the air gets warmer with height. This has the effect of making air on the surface very stable because cold air sinks and stable. Ice fog is present; this type of fog looks like diamond dust and exists only where it is very cold and dry. Even the slightest sounds carry for miles and moisture from ones breath freezes and falls from the air.

To Break an Inversion

Today an inversion often occurs in southern California (due to the cold water upwelling along the coast) with air surface temperatures of 50 degrees Fahrenheit and the temperatures at the top of the inversion several thousand feet up of 70 degrees Fahrenheit. In this example the air on the surface must warm 20 degrees Fahrenheit before the surface inversion is broken and the cloud layer clears and sunshine reaches the ground. On all too many days this does not happen.

A glacier has surface air at least a -40 degrees Fahrenheit and has a strong inversion aloft. Temperatures at the top of the inversion are anywhere from 0 to 50 degrees Fahrenheit, meaning it takes an increase of 40 to 90 degrees Fahrenheit to break the inversion. This rarely happens.

Due to this strong inversion over glaciers, sunshine alone from the short summers was not strong enough to break the icy grip of the ice. Only the return of the Jet Stream (when the Arctic Ocean finally ices over again) again brings strong warm westerly winds that melt the heart of the glaciers.

Air Masses

In a huge misconception much of what we consider to be weather such as cold fronts and storm, are not the cause of weather. No these weather patterns are just a reflection of the true cause of weather, which are air masses of one temperature moving into an area where air masses are of a different temperature. It is the interaction of air masses hundreds of miles across with different temperatures that cause frontal type weather with winds, clouds, and precipitation. Other causes exist but the movement of air masses cause most if not all of the world's frontal type weather above 20 degrees latitude (along the equator most of the air is hot and true air masses of different temperatures are largely not present). Air masses are often stationary next to one another, in which case you have a stationary frontal boundary between them.

An air mass can be defined as a large body of air with similar temperature and moisture properties throughout. A source region of an air mass is any area over which this air can stagnant, if it stays long enough it begins to take on the characteristics of the surface below. Air masses have specific physical properties, for example: cold air masses created over continents in the winter have higher pressure and lower temperatures towards the middle and rotate clockwise (in the northern hemisphere) at the surface, this results in cold air moving southward on the right side of the air mass and warmer air moving northward on the left side of the air mass. Shown by Graphic 13.

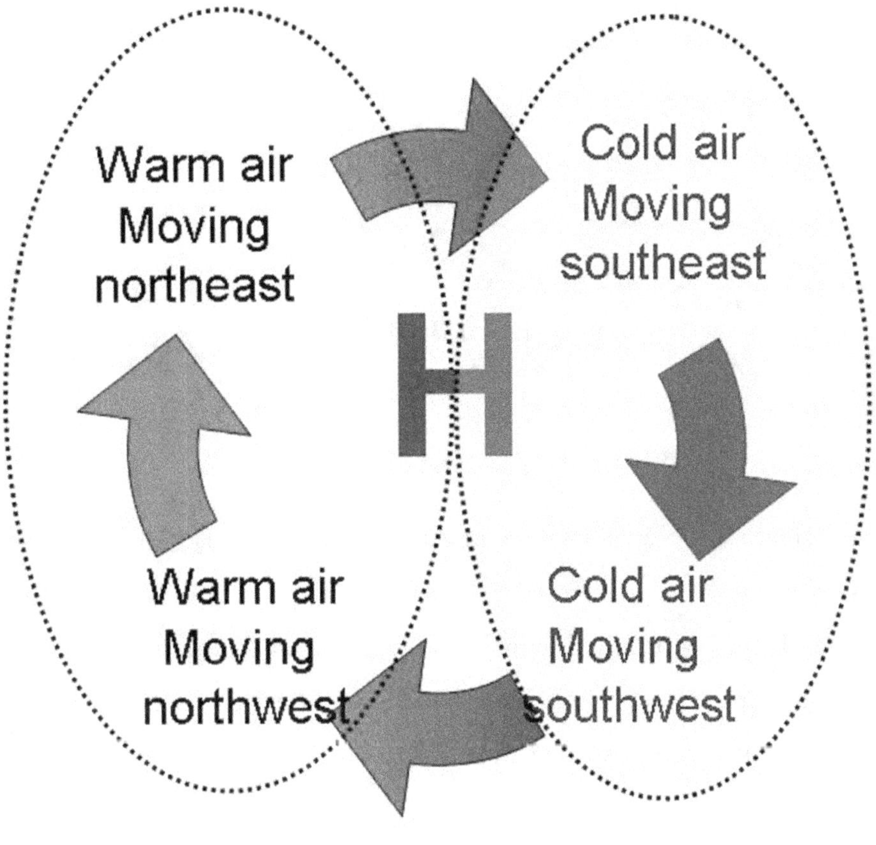

Graphic 13. Wind circulation around a High Pressure System

A cold High has clockwise rotation (in northern hemisphere) at the surface as shown by Graphic 13. This shows surface winds on the east side winds are towards the south and are cold winds (stable with a strong inversion), on the west side winds are from the south (unstable, the inversion is not as strong). The conditions above a surface cold core high air are often counter clockwise aloft as shown in Graphic 14.

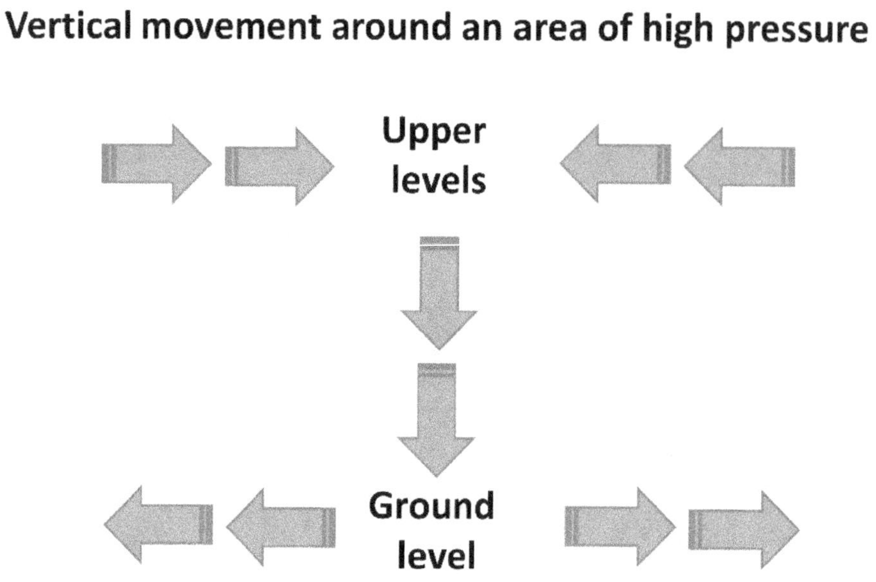

Graphic 14. Vertical wind circulation in an area of High Pressure

Warm core Lows reverse aloft, think convergence at low levels and divergence aloft. A real world example is a hurricane; storm force winds blow inward at the surface and reverse above 24,000 feet. Graphic 14 shows a cold high which is backwards from a hurricane. Like a reverse chimney effect – in at the top – out the bottom.

I digress AGAIN, a near fatal form of ADD I think.

As air masses move they interact with air masses of different temperatures and create fronts; cold air moving into warmer air is a cold front, warm air moving into colder air is a warm front. This is what we tend to think when we think of weather: fronts, rain, snow, thunderstorms, and wind.

In today's world

When a strong upper level disturbance goes over a stagnant air mass this often causes enough air to pile in behind the air mass that it begins to move. Thus it is disturbances in the world's upper level Jet Stream that cause much of the world's weather. As air masses move from their source regions, dynamics between conflicting air masses cause frontal type weather to happen.

In an Ice Age

During the last Ice Age extremely cold air masses formed over Eurasia, Greenland, and North America with temperatures down to minus 40 to 100 degrees Fahrenheit. This type of air mass circulates clockwise and results in extremely cold dry air blowing southward on the east side of these air masses.

Wherever one of these cold stagnate air masses faced a moisture source, such as the: Atlantic, Pacific, Arctic, Black Sea, Mediterranean Sea, or the Gulf Of Mexico (any large body of water) ice began to build, building a few feet or hundreds of feet a year (depending on the moisture source), gradually building to thousands of feet over thousands of years. This arrangement was effective because while on the earth's surface the cold winds blow out from the source, above the cold surface winds the winds often reverse. Resulting in warm moist air blowing from water sources towards the glaciers, right over the top of very cold surface air (resulting in freezing rain or snow). Off and on this slow building of ice lasted for over 100,000 years.

At the end of the last Ice Age when the Jet Stream finally returned to normal, the interior air masses over the glaciers slowly got warmer. This warming is helped by air rising up and over the glaciers top and down into the heart of the glaciers.

- Moist air that rises up the side of a mountain cools at 3.5 degrees Fahrenheit per 1,000 feet (the wet adiabatic lapse rate). This is the side where it rains.
- Once air reaches the top of a mountain and falls down the other side it warms at 5.5 degrees Fahrenheit per 1,000 feet (this is the dry adiabatic lapse rate).

As air flows up and over a 10,000 glacier, moist air that started as 50 degrees Fahrenheit cools at 3.5 degrees Fahrenheit per 1000 feet (moist adiabatic cooling). This results in 15 degree Fahrenheit air at the top, this now dry air descends on the other side and warms at 5.5 degrees Fahrenheit per 1000 feet (dry adiabatic warming); resulting in 70 degree temperature air at the bottom (adiabatic compression). This warming is why deserts are on the leeside of mountains. It helped melt the glaciers from the inside out, explaining why freshwater lakes formed in the interior.

Huge glaciers formed over Europe as far south as France and the mountains of Spain, stopping only on the northern side of the mountain chains separating the Mediterranean Sea from Europe. In Eurasia glaciers formed as far south as the Black Sea, these glaciers sucking moisture from both the Black Sea and Arctic Ocean to build as high as possible. In North America ice formed east of the Rocky Mountains and the Appalachians, as far north as the Arctic Ocean with ice fingers extending as far south as north Texas. They formed elsewhere also, but these were the biggies.

Lightning

Lightning does often strike the same place twice
It is why we have ground wires attached to crosses mounted on church roofs

If you hear thunder
it's a safe bet that lightning is present, time to get off both the roof and the fairway

CHAPTER 8

LEGENDS

Thousands of years before man recorded our history on papyrus or chiseled symbols on stone, our race kept oral records. These oral records were not legends and viewed with suspicion as they are today, rather they were the only way one generation passed critical knowledge to their children They were the words of our fathers and viewed with reverence. Those who carried these words were seen as carriers of sacred lore who took their duties on with deadly seriousness. In societies of the time, any twisting of these histories was heresy. If only it was so today, when information is twisted to fit any desired agenda.

The Sea People live on islands in the north eastern Indian Ocean and on Christmas Day 2004 when the sea withdrew far offshore these ignorant people who don't even have a written language immediately gathered all their loved ones and ran for higher ground. Folklore told them to do so. Tribes still following traditional ways lost no one to the deadly tsunami that killed so many with far greater education. Kind of makes me wonder who the ignorant really are?

The last big killer Tsunami in the Pacific was several hundred years ago occurring before the west coast of the United States became over populated, no telling how long ago the last big killer Tsunami happened in the Indian Ocean. It was before Europeans began capitalizing on the area 500 years ago. This is the minimum number of years that the Sea People had not personally seen an event like this occur and links in a chain, every story teller did a perfect job of passing critical information to succeeding generations ensuring words were repeated accurately and important concepts were passed down to the next generation. A long time ago a generation lost many loved ones to such an event and they did not want their descendants to suffer the same loss and passed critical information down to their children so they did not suffer the same fate, indeed a labor of love covering many generations.

10,000 year old under water structures in the East China Sea near Okinawa

Imagine if today the Pacific and Atlantic Ocean's started rising a foot every twenty years for 2,000 years; eventually rising 400 feet. It changes the very geography of the world and water would cover much of what man has created (at least up to 400 feet above sea level).

This huge change to the geography of the world happened 10,000 years ago; much of the history of the world became submerged, a tragic loss for mankind. During this period most dwelling were constructed from either stone or wood and we simply demolished dwelling about to be submerged and reconstructed them at higher elevations with these same materials.

In the Atlantic if the named list of hurricanes goes through the alphabet it is a rare event, in the West Pacific it is typical to get over 45 named storms a year. This is because the Pacific is much wider and the storms get more time to develop. These storms often grow to 1200 miles wide, huge monsters with eyes nearly two hundred miles across. Okinawa can be compared to Bermuda in the Atlantic, it is located in the middle of Typhoon Alley and every year this tiny island in the western Pacific is hit by multiple storms. These may be category one storms that barely affect normal life or massive category five storms that remove every bit of loose trash and debris from the island. It is a pretty clean place. It has been this way for hundreds of thousands of years, the Okinawa people hardly notice storms of any size passes over the island. If you make a wood house in Okinawa, it had better be strong. It is preferred to make concrete structures that weather storms with ease. Cinder blocks are not good as the winds blow water right threw them.

Okinawa is part of a chain of islands located on the eastern edge of the East China Sea. To the east the water drops off rapidly becoming thousands of feet deep. In contrast to the west of Okinawa the water is only 200 to 300 feet deep all the way to China and northwest to the Korean peninsula and the Yellow Sea. During the last ice age from about 10,000 to 70,000 years ago this entire sea was above sea level.

Today two styles of old castles prevail in Okinawa, one (Shuri Castle) can be viewed as somewhat a mix of both Chinese and Japanese and is fairly recent, it is very ornate with a large over hanging roof. During WWII the Japanese and Americans fought savagely over this castle, it was destroyed in the battle. My son's Okinawan grandfather was killed it this battle for the castle while his American grandfather was on a ship in the US fleet fighting the battle, his was struck by a Kamikaze (the third time). Another style is more ancient with low wide walls that have a slight inward slope and stand as high as 50 feet, these are constructed of stones. Dozens of examples of these older castle mounds exist on Okinawa, very common; they are hundreds to thousands of years old.

The most exciting archeological (and ignored) find since the pyramids are located off the coast of Okinawa, these are structures that appear to be predecessors of the ancient castles of Okinawa. Except these are located under 100 to 300 feet of water. Many large underwater structures and complexes have been found some of which may be old castles and others ceremonial areas. It does appear these are manmade and would be at least 10,000 years old. This makes Okinawa the site of the oldest known civilization, where a people lost in the mists of time were able to cut granite with precision and construct large typhoon proof structures that have lasted 10,000 years underwater. This area has always been viewed by Okinawan's with a huge amount of religious respect and awe, now we may know why.

10,000 years ago Okinawa was the coastline of a much larger China. The East China Sea to the west and the Yellow Sea to the northwest are both shallow seas just a few hundred feet deep, both of these shallow seas were dry 10,000 years ago. Where these artifacts are located due to difference in sea level; 10,000 years ago instead of this area being 300 feet underwater, it was 100 foot above sea level.

It has been just a decades that the mainstream science community has even considered that sea levels were hundreds of feet lower during Ice Ages. They just are not receptive to new ideas even when faced with evidence. Today there is reluctance by the scientific community to admit these structures near Okinawa even exist and are manmade that is preventing a much wider discussion of the possibility that Asia and Okinawa played a much larger role in history than previously thought. If these same underwater structures were found in the Aegean Sea, archeologist would be all over it. Prejudice comes in so many forms but always has an ugly face.

Okinawa's people are a kind and gentle people known for their longevity and patience. They do have some particularities such as being very laid back; for instance for staged events such as marriages and ceremonies they are on time, while for social activities like dinner, 7 o'clock means you can start thinking about getting ready at that time and show up around 9 or 10. They just don't like to hurry; of course this drives time conscious Americans crazy. Okinawa is very independent, even today the Japanese flag is not flown at schools or local government building as hard feeling from getting thrown under the wheels of the bus at the end of World War II remain as Okinawa was the site of the last large scale land and sea battle of World War II. Okinawa also has its own language, completely different from either Japanese or Chinese.

Perhaps these kind and gentle people who just want to be left alone are direct descendants of the people who constructed the castles that were buried under the waves of the sea 10,000 years ago. They have their own language, their own culture, are genetically different from either the Japanese or Chinese, and always wanted to be just be left alone. It is my belief that the Okinawa people as a whole would rather let their ancestors sleep.

Scientist today are reluctant to consider viewpoints that differs from accepted theory, thus a city of manmade origin is sitting on the bottom of the sea and few are taking notice. Herd instinct has taken over the scientific community and anyone who has theories even slightly different than the rest of the herd are viewed with suspicion. It is really sad that in the face of nearly irrefutable evidence that scientists refuse to admit that a civilization existed in Okinawa 10,000 years ago.

It is generally OK to build on accepted theory, but not OK to throw out accepted theory and start over. Albert Einstein did this very thing when he published his theory of relativity in 1901, yet he had to work in the patent office for 8 more years before his ideas were fully embraced by the scientific community. The bottom line is that it is difficult to get accepted, even if you are Albert Einstein. I wonder how many Einstein's (without Doctorate degrees) who have valid ideas and theories are laboring in a vacuum, with no chance of ever becoming accepted or even noticed. I am no Einstein but this makes me angry.

Remember China was free of ice, from the Himalayas in the south to the mountain chains to the north were not iced over during the last Ice Age. People living in Asia during the last Ice Age had 70,000 to 100,000 years of warmth and rain, a perfect place for a civilization to exist. In addition for most of that time they had a perfectly good land bridge to Alaska into North America.

Today if our oceans rose 400 feet most all of our coastal cities go under water; all of Florida and most of other coastal states would submerge. Now imagine our forefathers 10,000 years ago who lost everything and many loved ones to a global rise in sea level of 400 feet in 2,000 years (a foot every 5 years). The survivors wanted to pass these memories to their children. Imagine a hundred generations of man living on the sea shores of an ancient world watching dwelling and villages gradually get lost to a slowly and continually rising sea. These people used verbal memories as their vehicle to pass warnings down to succeeding generations, in the same manner of the previous generations of Sea People who had endured tsunami on too many occasions. Our forefathers did this for us when they warned us of the huge floods, an event that occurred 8,000 to 12,000 years ago.

During an Ice Age the Mediterranean Sea evaporates more than it gains

The Mediterranean Sea is continually trying to evaporate away and loses approximately an inch of water every day to evaporation. Not much you say? Consider the net loss of water every day, water that is replaced by water coming in from the Atlantic. The sill of the Straits of Gibraltar is 930 feet deep, decrease the Atlantic sea level by 400 feet and the sill of the Straits of Gibraltar now becomes 530 feet.

Due to diminished sea levels in the ice age the Mediterranean became starved of water. Two ocean straits the Gibraltar at the western end of and the Dardanelles at the eastern end had a huge impact on humanities growth: this is due to the shallowness of these straits. The Straits of Gibraltar is 930 feet deep while the Dardanelles (located between the Mediterranean and the Black Sea) is less than 200 feet deep. In the case of the Straits of Gibraltar which is deepest in the middle and slopes upwards towards the sides, a global sea level drop of 400 feet decreases the volume of water that can flow over the sill by over half.

This severely limited the amount of water that could flow from the Atlantic to the Mediterranean. When sea levels dropped 400 feet both the Mediterranean and Black Seas become far shallower than either the Atlantic or Pacific Oceans.

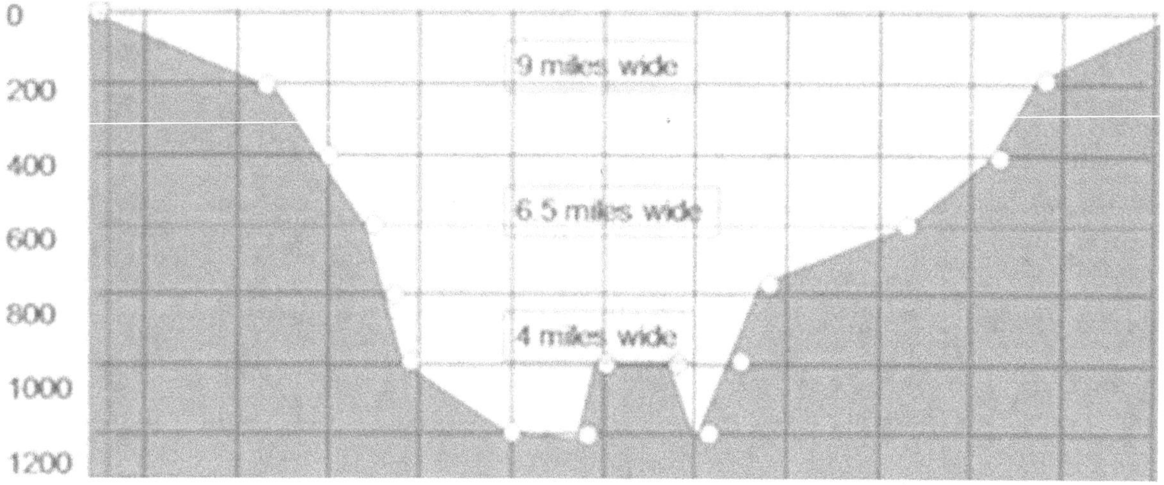

Graphic 15. Camarinal Sill at the entrance of the Straits of Gibraltar, Depth in Feet verses Width in miles

The Camarinal Sill at the entrance of the Straits of Gibraltar into the Mediterranean is an average of 930 feet deep and about 11 miles wide and shaped like a speed boats V shaped hull (shown by Graphic 15), deepest in the center and slopping upwards at 45 degree angles. At the end of the last Ice Age this sill became 530 foot deep and four miles narrower. This resulted in a significant reduction in the volume of water that flowed through the strait, by over half.

Interestingly in the 1920's it was proposed by a German engineer named Herman Sorgel to put a hydroelectric dam in the Straits of Gibraltar and let the Mediterranean Sea evaporate several hundred meters, then use the inflow from the Atlantic to generate electricity. It also opens up a lot of real estate that is now under water. Politically he was out of his mind (he was German after all), but the concept is sound. This happened naturally during the Ice Ages when the world's sea levels fell 400 feet.

A 400 foot drop in the world's oceans does not sound like a great deal until one considers that water flowing in from the Atlantic through the Straits of Gibraltar is the only significant source of water for the Mediterranean Sea. This inflow helps to keep the Mediterranean at the same level as the rest of the world's oceans; evaporation is balanced by inward flow:

Mediterranean Surface area:	970,000 square miles
Evaporation per Square Mile at 1 inch per day:	17 million gallons
Evaporation total in all Mediterranean per day:	16,000,000,000,000 gallons

If just 1/100 of this water is not replaced the Mediterranean loses a foot every four years, if this condition persists for 1,000 years, this is a loss of 250 feet, for 10,000 years a loss of 2,500 feet. The bottom line is at the height of the ice age it takes a lot more water than is capable of flowing over the Camarinal Sill to keep the Mediterranean from evaporating away. Due to differences in topography of the sea floor this lowering of the Mediterranean is not uniform; sea levels in the Mediterranean Sea decreased hundreds of feet in the western area, while the Aegean Sea at the eastern end decreased as much as two thousand feet.

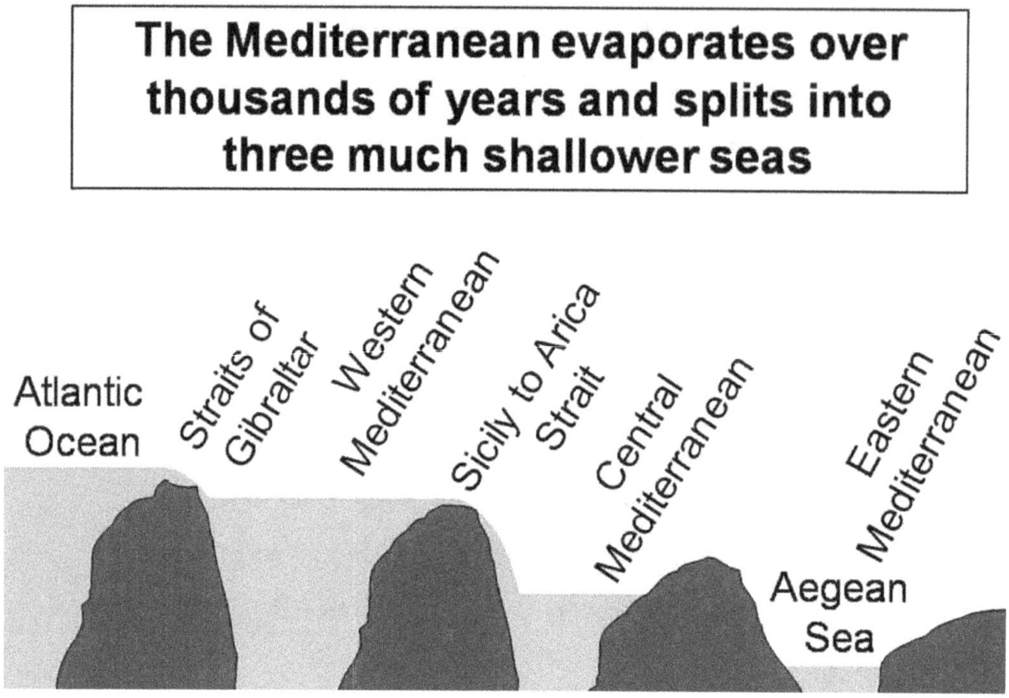

Graphic 16. Sea level changes in Mediterranean Sea during Ice Ages

Sea level drops in the Mediterranean are greater the farther away one gets from the straits of Gibraltar. This is shown in Graphic 16. This difference in sea level is even more pronounced because the Mediterranean splits into three distinct seas; for a sea level drop of 400 foot in the Atlantic the western portion of the Mediterranean drops between 400 to 800 feet, the middle of the Mediterranean south of Sicily drops between 500 and 1,000 feet, and the area of the Aegean Sea drops several thousand feet. This is where we need to look for traces of habitation by humans 10,000 years ago – deep water – Aegean Sea.

Today they have found quite a few underwater ruins in the Mediterranean, but they have all been in shallow water usually visible from the surface. They need to go deeper in the Aegean and check sea mounts.

Eden, Dragons, Legends, and the Sea

The time between man using solely verbal records and shifting to written records is hidden far back in the dusts of time. In this transition many legends that were simply unbelievable were disregarded. Granted in a world where verbal history is the sole source of information it was almost impossible to separate fact from fiction. I expect that sea stories that travelers told at the time were really wild. When writing was invented these verbal record keepers had a major problem convincing the writers of history exactly what was bull and what was important.

Imagine if we had no written language and were trying to pass down our children the horrors of the 2004 Tsunami but had no video, written words, or pictures, that we were limited solely to verbal stories from generation to generation. How many generations does it take our children to think Grand Pa was simply out of his mind, waves never get that big and its impossible for so many people to just be swept away by the sea, Or for small changes in the story to occur as it passed from generation to generation and gradually morph into something that did not really describe the event at all. Unless a great deal of care is taken by the chain of story tellers (like the Sea People did) this is liable to happen every time a story is told and retold by storytellers.

I am not an expert on the bible but it is painfully obvious to me that it is a study book for teaching morals to succeeding generations. The multitude of passages and endless collection of parables are constructed as tools to teach almost every possible aspect of life and living. The bible contains accumulated knowledge from those years before writing. This is obvious by the length of the passages, as almost all of them are short. This was when lessons had to be told and retold to ensure that knowledge was not lost. They had to be short, sweet and to the point; otherwise they could not be absorbed in one sitting. I think a lot of the events in the bible actually happened but were modified to fit the agenda of a rising church. Much like far right or left wing commenters do today when describing world events. It's as if Fox news went back 2000 years and wrote the bible.

My take (and a twisted one) is that events described in the bible concerning Adam, Eve, an apple and a serpent were lessons to teach virtues of many types. The apple (email, person, or concept) who comes from the outside of the group may look perfectly wholesome and benign, yet be malignant and rotten to the core (read traitor, spam, or email virus here), accepting him or her (or opening infected email) was a perfect recipes for destruction. Adam and Eve certainly were not eating from the tree of knowledge or opening email, maybe eating passion fruit, or most likely just passion and she was playing with a serpent. This is clearly a multipurpose lesson that can be used in many situations as can most other passages of the bible.

A flood clearly happened (multiple ones) and the description of 40 days of rain prior to the flood can easily be written off as a later attempt by story tellers to rationalize the sea rising and taking everything and everyone away, an event that occurred when it was not raining. It didn't make sense to them and they didn't want to argue about it at every storytelling, so they said it rained. The obvious cause of this rise in ocean levels is the melting of the glaciers and ice at the end of the last Ice Age. It simply started as 40 days of flood and morphed into 40 days of rain because writers of the bible did not understand what had happened. Once the Black Sea filled and the Aegean started filling in earnest, this may have been how long it took for the main islands of the Aegean Sea to flood.

Dragons

In another example how verbal history is been passed down to later generations, in 1991 I witnessed the volcanic eruption of Mount Pinatubo from 21 miles away. On numerous occasions the eruptive column reached the stratosphere, this is when the ash mushrooms outward (like a nuclear detonation) in all directions. A very scary experience and in many different ways: I saw it, heard it, smelled it, and could feel the magma moving deep underground as continuous harmonic tremors (magma moving underground) shook the earth, and a few larger quakes that made me run for doorways. Experiencing an ash falls is like a heavy snow storm except sand is falling from the sky and it's dark and quiet. Falling ash creates lots of static and near constant weird green lightning was striking everywhere that devastated the surrounding forest. You see the lightning striking very close but hear almost nothing.

On three consecutive days I awoke at high noon (was working nights) to a pitch black day and falling ash. During the night we watched erupt many times on our weather radar, in just a few minutes after the start of an eruption it went straight up for 85,000 feet with no outward flare at all; like laying a pencil vertically on the radar screen. Our radar topped out at 85,000, the columns went a lot higher. The movement of magma in the magma column created lots of static and it was illuminated by thousands of greenish lightning bolts snaking over the huge mass. Like watching a huge black thunderstorm with ominous green lightning illuminating it. Even the mushroom cloud that expanded outward was lit up by this weird lightning, snaking around on the base of the cloud. All in all it was a nightmare.

An earthquake when you are driving feels like you're in an accident. The car is shaking and you look around angrily wondering who is running into your car. An eruption is accompanied by a lot of them.

It's almost hypnotic to see a magma column up close. You know you are in a big eruption because you start watching it; keep watching it, then your neck starts hurting like hell because you are looking straight up at the edge of a huge mushroom cloud. My neck hurt for weeks after it was done.

One day while watching a big eruption with its black mushroom cloud (a layer at least 100,000 feet up) expanding over me and covering the sky, it got dark. I saw hundreds of lightning spikes snaking across the bottom of the cloud and I realized it looked exactly like a dragon was flying in the cloud and blowing out a big stream of fire across its bottom surface. Little wonder the ancients had legends of dragons spewing fire across the sky. I might add that even in this day and age such an event is terrifying, it was not hard imagining a dragon flying across the sky; spewing fire and death.

As we explain legends of dragons by volcanic eruptions, we can also explain the existence of hundreds of legends (from many different cultures) that speak of floods. With so many flood legends, it would be a logical deduction that only one big one covered the entire world. The Bible speaks of Noah and recounts a flood in which only Noah's family and two of every creature survived. Perhaps this is a true story, at least from the viewpoint of some who survived. It did flood their world. An event that happened 10,000 years ago. Over thousands of years it just morphed it a bit from the original version. The writers of the bible

In the years before we learned to write all of our learning was done by word of mouth and song. We were like dry sponges soaking up whatever knowledge we could, storytelling was the movies of the day. I believe the bible is a compilation of thousands of these stories and sayings (gospel), that both taught a moral lesson and kept you entertained. A double bonus for the teachers, you can teach someone something and get dinner.

In the depths of the last Ice Age we learned to use fire, with it we could survive living in the shadows of the towering masses of ice in the north. We probably first encountered fire from natural seeps of natural gas that were static and then burned for thousands of years. This at a time when man was dying in the cold realms of the day and every moment a challenge to just survive. In a world of cold finding water to drink was a nearly insurmountable challenge. With fire we could finally ward off the darkness of the night, melt ice to drink, and cook our food. It put us on equal ground with the carnivores. The day we conquered fire we climbed from the bottom to the top of the food chain and finally able to keep the terrors of the night at bay. Plus barbeque was invented (if they only had beer it would have been perfect, "pass me a bud light and another one of those mammoth ribs" :).

Imagine a world where ice has driven most bands of humans far from the northern regions. Some of them hit the jackpot as they found a paradise, as gradual drops in sea levels created a series of lush near tropical islands in the Aegean Sea with no natural predators. These would appear over generations and get bigger and more attractive as time went on. For humans who were accustomed to routinely battling hungry four footed carnivores of all sorts, this lack of danger certainly fit the biblical version of the Garden of Eden. A place of plenty, safe for children to live and grow; where the only danger is knowledge itself.

How much water became locked into Ice during the Ice Age?

The moisture that grew the glaciers came from the world's oceans, today they cover 129,671,000 square miles of the planet's surface. 17,000 years ago sea levels were four hundred feet lower than today (400 feet is 1/15 of a nautical mile = 6,000 feet); dividing todays water coverage by 15 equals the amount of water locked in ice. 8,000,000 square miles. This is a huge amount over

13,968,990,016,400,000,000 gallons. Over 167,627,880,196,800,000,000 twelve ounce bottles of water. If you lined them up end to end, enough bottles of water to go around the world 600 billion times.

What is incredible is what took 110,000 years to freeze took only 5,500 years to melt, starting about 17,000 years ago and ending about 10,000 years ago. Flooding was rampant as eight million square miles of water sought its way to the sea. Northeast of the Aegean Sea is a huge watershed that drains into the Black Sea where approximately a quarter of the melt water from the melting ice drained (over 22,000,000 gallons a second for 5,000 years) and when that basin filled to the level of the dry Dardanelles Straits (located between the Aegean and the Black Seas) water began flowing into the Aegean Sea in earnest. It also flowed into the Mediterranean (in a much greater volume) from the Atlantic via the Straits of Gibraltar. This flood lasted at least 5,000 years with an average of 44,000,000 gallons a second first filling the Aegean basin and then the rest of the Mediterranean. This is a LOT of water, a FLOOD.

Prior to the flood, the Aegean Sea was a paradise with sea levels several thousand feet lower than today with many islands and land masses exposed for use by man. Islands free of the large predators, safe havens for man to live and thrive for over 60,000 years. It was our cradle, at least for a portion of mankind.

Today some of the largest thunderstorms over North America happen over the Grand Canyon, when moisture from the Pacific flows northeastward into Arizona from Baja during August and September, these are enormous and scary to witness as they extend from the bottom of the Grand Canyon to well over 60,000 feet. When the Mediterranean Sea dropped several thousand feet in its eastern region / Aegean Sea – storms of this magnitude or greater happened almost every day. These storms pumped moisture far up into the atmosphere for thousands of years bringing life to the deserts of North Africa and the Arabian Peninsula, creating lush forests where now there is only sand and fossils.

The moisture from the Nile waterfall also created a thick maritime layer of stratus and stratocumulus clouds in the Mediterranean basin extending far north. This created a climate similar to today's coastal Southern California, very cool and comfortable to live in. This maritime layer was instrumental when Toba blew up 75,000 years ago and disrupted the world's ozone layer, as this layer protected this paradise from the sun's harmful ultraviolet rays that were deadly to both animals and plant life.

The Nile river falling hundreds of feet as it poured into the southern Aegean Sea was an awesome waterfall and like present day Niagara Falls, this waterfall eventually cut a deep canyon hundreds of feet deep and many miles upriver. The deep canyon of the Nile and its fast flowing river was a very effective barrier and prevented man from traveling east to west in this area, while the cliffs at the northern coast of Africa prevented travel from north to south. This separated societies in the area and why populations in the area are so different from each other.

After the ice age the Nile filled up with gravel and was a major problem for engineers of the Aswan dam as they has to deal with hundreds of feet of gravel below the existing riverbed. Now imagine Niagara Falls with a drop of 2,000 feet. This is the Nile waterfall, an amazing sight and created conditions conducive to the evaporation of water. These falls pumped huge amounts of moisture directly into the lower atmosphere.

Even today the Nile River is a monster, but at the height of the Ice Age the daily flow into the eastern basin and southern Aegean Sea was much more than today. This was the primary source of fresh water for the Aegean Sea and made it the least salty of the three Mediterranean basins. The falls brought fish and debris in the water and carrion eaters of both the sea and land made its shores home. Fishing was possible as well, but why fish when food falls from the sky? At the base of the falls many predators coexisted side by side, due to the plentiful food of the area.

In this area lives the bearded buzzard that likes to eat bone marrow. It waits until other animals have stripped everything else from the bodies of carrion then picks up these bones and flies high over rocky terrain and drops the bones onto rocks (in recent times there are unconfirmed reports that they've hit bald headed men on the head). The bones crack open revealing the marrow and the bird feasts on his reward. They also like turtles (and other small prey) and kill them the same way. This is a learned behavior and takes young birds years to learn. Is this buzzard an ancestor of the thousands that once feasted on abundant carrion at the base of the Nile Falls through the Ice Age's? I think so, these ancestors of theirs had learned that carrion falling a thousand feed was choice fare and if it wasn't dead or tenderized enough, they just flew it up and dropped it again. When climate changed they modified their behavior to fit their new environment. I may be wrong but in any case is a very interesting bird.

Eden

The extensive drop in sea levels created a paradise. Then disaster struck. At the end of the Ice Age when weather patterns became much like they are today, the ice started melting and what took 100,000 years to freeze; melted in just 6,000 years creating flooding on a biblical scale.

- In Oregon the Columbia River Gorge was created when melt water built up over many years behind the huge walls of melting glaciers. This water eventually found its way out through cracks in the ice, these enlarged in size and the water rushed out. Afterwards the cracks refroze and the process was repeated several times (over hundreds of years) until the walls were fully breached and continuous flow occurred.
- In Eurasia, the melt water feed into the Black Sea and after that basin was full, onward into the Mediterranean Sea causing the eastern Mediterranean Sea to fill from both the east and west.
- The Mediterranean Sea was deluged in all directions by melting ice, causing it to rise at a far greater rate than the larger oceans.
- The Aegean was at a lower level than the rest of the oceans of the world and filled faster.

The Pacific and Atlantic Oceans took 5,000 years to return to levels close to what we see today (they continue to rise), rising a foot every 20 years (not enough to be noticed) while the Mediterranean Sea by contrast rose much faster, as much as a foot every month. The Aegean Sea rose even faster, up to five feet a day especially in the first years of the flood.

Static

Most accidents that happen with static electricity – happen in a clear blue sky. This is when dry hair and dry air mix best to create static. This is when static electricity is unable to bleed off naturally into moist air and objects such as hair charge up naturally.

CHAPTER 9

THE ICE RECORD

Ice cores taken from the oldest glaciers of Antarctica and Greenland give us a bird's eye view into the climate of the last 800,000 years (400,000 years for Greenland). When scientists analyze the ice they use special razors and scrape off a thin layer at a time and analyze it for gases and dust. They can determine the composition and temperature of the air when the ice was created. These ice cores give us a virtual thermometer that shows the temperature of the earth's atmosphere over thousands of years. Looking at the last Ice Age period taken from Greenland glaciers in Graphic 17 (a repeat of Graphic 2), we can see temperature trends for the last 133,000 years.

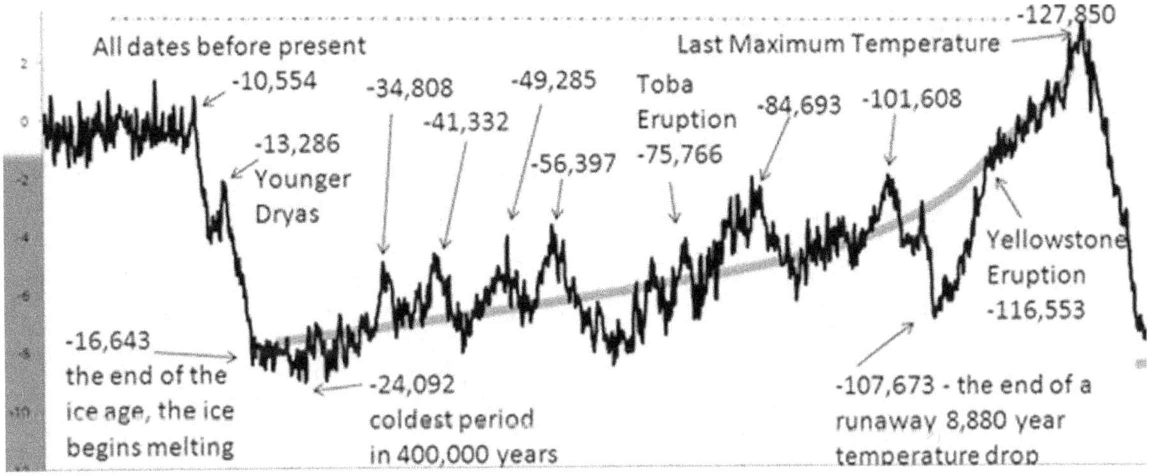

Graphic 17. 133,000 Year Temperature Trend (a repeat of graphic 2)

Graphic 17 shows the last 133,000 years and with smoothing it is easier to see a pattern does exist. A previous graphic, the Ice Record in Graphic 3 shows 400,000 years of temperatures and clearly shows a repeating pattern of warming followed by approximately 100,000 years of gradual cooling and warming. Like clockwork this pattern repeated four times in the last 400,000 years. We see a repeating saw tooth pattern of extreme warming followed by long term cooling that lasts on the average 100,000 years, where it gradually got colder and colder until all of a sudden it gets very warm again and the cycle repeats.

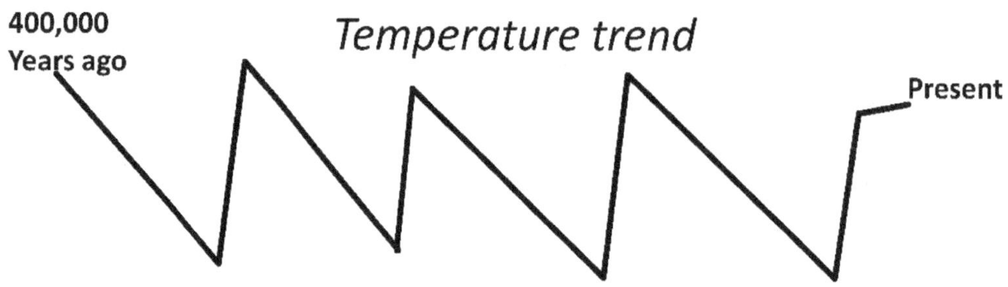

Graphic 18. Saw tooth pattern of the smoothed Ice Record for the last 400,000 years

Looking at Graphic 18, each of these zig's takes about 100,000 years and it is obvious that different events cause the beginning and ending of Ice Ages. We are looking both for a cooling process that takes 100,000 years and a trigger that stops the cooling and causes abrupt heating. Graphic 19 shows how I smoothed the record to order to get a clearer understanding of these events.

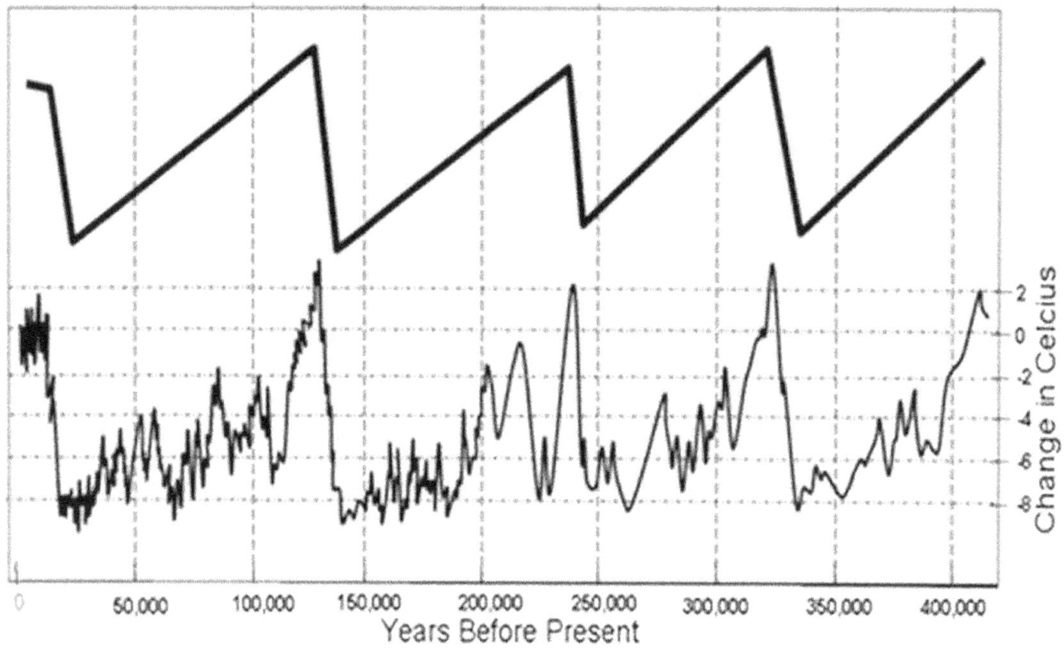

Graphic 19. Smoothing the Ice Record

Smoothing the Ice Record was not pretty as I ignored short term changes caused by either volcanic eruptions such as the eruption of Yellowstone 116,000 years ago resulting in a large temperature drop and short term temperature spikes caused by repeated Methane Hydrate releases. Instead, to create a trend I averaged the temperatures and came up with a fairly smooth curve. It's ugly, imprecise, and makes perfect sense as I am looking for the general trend.

It is the last 10,000 years that are very interesting, temperatures flattened out when they should have been rising. I argue that this is the result of man's rise after we mastered the use of fire. We have been modifying the weather for 10,000 years. It's not the date we started using fire as that was long before this, rather it's the date that we began using it wholesale as a tool to master the environment.

Winter

If it is winter, the sky is beautiful and it is also very warm

Surface winds are very breezy out of the south

Great kite flying weather

Check the weather channel

A cold front is on the way

postpone trips, and prepare for bad weather

CHAPTER 10

WHAT DO WE KNOW?

We know extreme warming starts the Ice Age Cycle and is followed by a period of extreme cooling. The most obvious culprit for starting this cycle is the melting of the Polar Ice Cap. The Ice Age Cycle typically takes 100,000 years and is comprised of a long continuous cooling trend, getting ever colder, followed by a period of rapid warming (sorry, I realize I am beating a dead horse, just trying to put it together). Some observations:

1. The Arctic Ocean was free of ice during the last Ice Age.
2. That methane spikes from the near explosive degassing of huge methane hydrate deposits on the worlds continental shelves are the cause of multiple warming spikes. These are shown on Graphic 17. These warming periods last several thousand years and then cooling takes back over. This is shown time and time again over the last 70,000 years, after the worlds sea levels had fell hundreds of feet.
3. That huge glaciers several miles thick covered large portions of the continents, holding over 8,000,000 square miles of water.
4. That several large volcanic eruptions occurred during the last Ice Age Cycle, the major ones are Toba 75,000 years ago and Yellowstone 116,000 years ago.
5. Due to the additional weight of glaciers over Greenland and Europe, the ocean floor in the 900 foot deep Icelandic Straits between the Atlantic and Arctic oceans rose at least 100 to 300 feet during the last half of the last Ice Age.

It seems as if God threw a switch at the end of the most recent Ice Age cycle (about 17,000 years ago) as one day it abruptly started warming and the ice gradually melted and for 6,000 years sea levels rose. This process continues to this day, the seas are still rising but much slower.

The combination of lowering global sea levels and rising ocean floor in the Icelandic Straits reduced water flow between the Atlantic and Arctic Oceans to its lowest amounts about 17,000 years ago. For a period of time (until the glaciers melted) the inward flow of water from the Atlantic was reduced to a fraction of what it was at the height of the Ice Age. The Iceland Straits became less than 400 feet deep. This reduced the flow of warm water into the Arctic to such a great extent that the Arctic Ocean froze over again and the current warm era started.

What is really questionable is how much did the Sea Levels drop? Sea Levels initially fell around 200 feet over 20,000 years; about 1 inch every 12 years; the Bering Straits become a land bridge approximately 105,000 years ago. Sea Levels then gradually dropped an additional 200 feet over the next 80,000 years. I base this on the temperature minimum that occurred 105,000 years ago, when the world warmed for a brief period. It is this period that the runaway fall in temperatures ceased and a different period started, when the climate repeatedly warmed and cooled in a 7,000 to 15,000 year cycle (still cooling overall). Another temperature minimum occurred 64,000 years ago.

Several possible explanations for this change:

1. That the Bering Straits became a land bridge 105,000 years ago and warm water from the Pacific quit flowing into the Arctic.
2. That the eruption of Yellowstone 116,000 years ago was the main cause of the runaway drop in temperature and these affects stopped 105,000 years ago, returning the world to the gradual downward trend it was in before.
3. This is a result of repeated degassing of huge Methane Hydrate deposits on the ocean floor that warmed the climate on multiple occasions.
4. Most likely a combination of all three.

Hurricanes

The first sign of a hurricane is an ever thickening layer of cirrus stratus clouds, for a day or two you will have few low clouds and the sun will have a large ring around it.

As the storm nears – the sun will increasing look as if you are looking through frosted glass. Then the sun gradually disappears as the storm nears
and the cloud layers get lower and lower.

If by chance the storm comes ashore during high tide, the storm surge may be many feet over what is normal high tide.
Add 10 to 20 feet on top of the regular tide – more for a category 4 or 5.

This storm surge will be located where the winds blow on-shore, in the northern hemisphere this is on the right hand side

CHAPTER 11

THE JET STREAM

When I first wrote Tipping Point in 2010 I believed the Jet Stream was still healthy and surely last for ten more years. I was wrong. It is now 2015 and many climatic changes we see happening today are directly related to a weaker and less defined Jet Stream. We have even started calling it another name, climate change instead of climate warming. I hate it when I am right, especially when no one listens.

Think of air as the blood of the world's atmosphere and upper level Jet Streams are the circulation system. Just like our own body, a healthy circulation system is critical to the health of the planet's atmosphere. Jet Streams are part of a worldwide atmospheric circulation system that distributes moisture and heat across the globe and are critical to moderating the climate and preventing localized extremes of both temperature and moisture. The Jet Stream is a major component of this circulation system and helps transport air masses (with all the differences inherent within these air masses, such as variations in pressure, temperature, and moisture) from one region to another.

Living on the surface of the planet, we do not have an understanding of the meandering rivers of air that lay far above our heads. We did not even know they existed until airplanes started flying above 25,000 feet. These Jet Streams are rotating rivers of air that flow high above the borders between cold and warm air masses. Like corkscrews, they rotate counterclockwise as they move forward resulting in clouds forming on the right upward motion side of the corkscrew and clear air on the downward side; these physical characteristics allow them to clearly show up on satellite.

Upper level Jet Streams are a reflection of temperature differences on the surface. The strongest of these Jet Streams is behind and far above the fastest moving surface cold fronts. These cold fronts typically separate cold dry continental polar air from warmer maritime tropical air over the oceans.
These jets are not contiguous, rather they extend in a series of segments all around the planet. They are most evident and strongest in the winter along the eastern sides of continents where the temperature differences between the continental polar air and maritime warm air masses are greatest.

The winds in Jet Stream segments are typically over a hundred and up to two hundred miles an hour. The Jet Stream is a major reason why we experience weather as we do, as variations in the direction and speed in the jets flow are a major reason why air masses move as they do. These variations cause disturbances (jet maximums) and do a wonderful job of piling up air behind stationary air masses, gradually pushing these air masses out of their source regions and into other areas. The Jet Stream normally moves west to east in waves with an amplitude of hundreds to thousands of miles. The world has an average of five to seven large waves in the Jet Stream around the world, with many smaller waves embedded within these large waves.

After the tipping point, an upper level ridge of High Pressure develops over the Arctic Ocean (over the warm core Surface Low), this is anti-cyclonic and works against the Jet Stream. Temperatures are coldest over the continents and this messes with the Jet Stream, breaking it into pieces, and preventing it from traveling its normal path.

When the Arctic becomes ice free and ice starts building over the continents the Jet Stream breaks into multiple regional Jets: one segment rotating around the cold Eurasian air mass over Asia, another segment rotating around the cold Canadian air mass over Canada, and a third segment rotating around a cold European air mass over Europe. These Jets rotate in circles around these intensely cold and dense air masses. A weak Jet Stream may still go around the world, but is not strong enough to cause the High Pressure systems over the continents to shift position. This relatively small change yields large results and we see a situation where air masses remain locked in place for long periods of time.

The world still receives the same amount of energy from the sun but instead of this energy being fairly distributed across the globe, a much higher percentage ends up being absorbed by the oceans and the air masses over them. Ocean currents carry this heat around the globe and into the Arctic Ocean by way of warm currents flowing through the Icelandic Straits and between Alaska and Asia.

CHAPTER 12

THE ICE COVERED ARCTIC OCEAN

It is difficult to comprehend how the Jet Stream affects climate, being something that we can't see. Even more difficult is to understand how the Arctic plays a crucial role in determining the world's weather. Like a small town it is all inter related and if everything works all is well; but let one thing break and the whole thing quickly goes down the crapper.

Also difficult to understand is how variations in temperature, density, pressure, and salinity create layers in the ocean and affect the climate (or vice versa). For instance prevailing winds cause upwelling currents along the west coast of the US bringing up vital nutrients for plant life, this is the base of the food chain for aquatic life. This current also modifies the overlying air mass and makes our coastal climate mild compared to the inner mountains and deserts.

Over the last few years a decrease in upwelling for the currents off the coast of Chile is driving the sardines away (nutrients in the water decrease), an industry that has lasted generations. Changes in atmospheric conditions play a large role in creating these currents and these in turn have an effect on the overlying air mass. The Arctic is special because climate changes there affect global weather patterns.

Like some people, sea water molecules are surprisingly prejudiced and they only like to associate with other water molecules of the same temperature, salinity and density. In the ocean this creates layers of water with different salinity and temperature. In the oceans you typically you have a mixing layer, this is the warm surface water churned by winds that may go down 10 to 20 feet and generally is all the same temperature (with a slight increase at the surface due to solar heating). Below this mixing layer the temperature gradually decreases to the bottom. Salinity is also highest at the surface due to evaporation, this increases the density of the water as highly salty water is heavier than less salty water.

The colder fresh water gets, the denser it becomes until it hits 40 Fahrenheit (sea water is lower due to the salt and resulting lower freezing point, 28.4 Fahrenheit). As temperatures cool below 40 F the water becomes lighter due to the formation of ice crystals (that are lighter than liquid water). As one nears the freezing point the number of ice crystals make up the majority and water is at its lightest point.

Thanks god that ice floats or else it would sink to the bottom of the ocean and eventually the entire ocean would to frozen top to bottom. Due to this quirk of floating ice, the temperature structure of the Arctic is completely ass backwards from warmer oceans.

In the Arctic; obviously ice floats and cools the top layer (like a cup of water with a thin layer of ice cubes on top). It is ice covered and coldest at the surface due to the floating ice. This cools the top few hundreds of feet, this depth varies according: currents, surface mixing, amount of evaporation, salinity of the water, speed of the winds, and temperature of the air. This is a static layer that has few vertical currents. Below this layer is the Atlantic middle water, a thick layer of warmer water that has come into the Arctic from the Atlantic in subsurface currents.

If evaporation occurs over open water in the Arctic a thin layer of surface water gets highly saline as the salt accumulates in the evaporation layer; big globs of this layer eventually get so heavy that it sinks to the bottom. This is the Arctic cold bottom water.

Like ice floating in a glass of water, surface water in the Arctic Ocean is at or near freezing, but as one descends downward both the temperature and density of the water gradually increases until the temperature is above 40 degrees Fahrenheit. This is the middle water in the Arctic a thick layer of warmer water that has come into the Arctic via the Greenland Straits, about half of the Arctic Ocean is this warm Atlantic water. Below this lays another layer of very cold dense water that extends to the bottom.

In the winter over an ice covered Arctic Ocean, the surface water under the ice gets colder as winds slowly cool the ice during the long Arctic night. A lack of vertical movement creates a very stable layer of very cold water over a warmer layer that has been pushed as deep as possible. Currents are weak between these layers.

An ice covered Arctic Ocean does a poor job of transferring moisture to the atmosphere and creates a dense layer of stable cold dry air on the surface of the ice pack. This Arctic air mass is not nearly as cold as the Siberian, Eurasian, or Canadian continental air masses; principally because land cools and heats much more rapidly than water. The air mass sitting over the Arctic Ice pack is not a strong High or Low, rather a series of weak Lows and Highs. What is true is the overall air mass of an ice covered Artic is much colder (and more stable) than an air mass over an open water Arctic.

Today the (ice covered) Arctic air mass does not interact directly with the Jet Stream that lays far to the south but does create a near contiguous area of very cold air between the winter Eurasian, Siberian, and Canadian Highs. This helps to create a nearly continuous Jet Stream that goes all around the world. The Arctic in effect acts as a virtual extension of the cold continental air masses creating a nearly continuous cap of cold air over the top of the world.

A 747 flying from Tokyo to San Francisco in the summer can take 14 hours, this is when the Jet Steam is very weak. That same flight might take 10 hours in the winter when the Jet Stream is strong and the 747 gets a tail wind. This is entirely a result of the temperature gradient between the Arctic and warmer regions to the south.

If the Jet Stream were to flow in a straight line weather would not happen as we know it and we would see few frontal systems moving across the hemisphere. Instead the Jet Stream moves in a series of upper level waves, like a sidewinder rattlesnake that is moving forward.

The Jet Stream moderates global temperatures. As these upper level rivers of fast moving air make their way across the continents, air piles up in front of them that decreases the stability of the Siberian, Eurasian, and Canadian air masses and in many cases end up kicking these air masses out of their source regions. This occasionally movement of continental air masses out of source regions keeps them from stagnating too long and becoming either too cold or too hot.

Driving in Fog (sand, smoke, snow and rain)

If you are on a fast road and you are unable to see
You have driven into an extreme life threatening situation
Stop a long way before you are unable to see
That way the people behind you can see you and don't run into you
It's best to watch the news and weather forecast and don't get into this situation

CHAPTER 13

THE ICE FREE ARCTIC OCEAN

During the last Ice Age the Arctic was free of ice. How could this happen, especially with huge glacier covered land masses surrounding the Arctic Ocean?

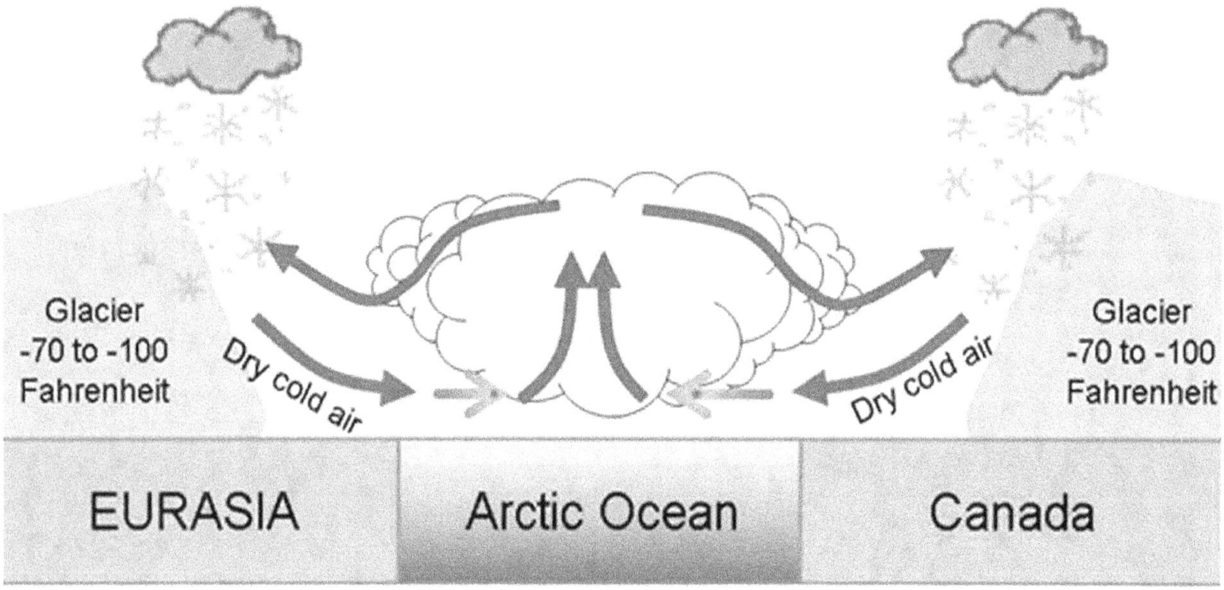

Graphic 20. Arctic Ocean wind circulation and Glacier Building over the Continents

It is like two mountain peaks sitting next to each other that has a valley (with lower elevation) between them. The atmosphere works in much the same way and adjacent areas of High Pressure have an area of Lower pressure between them. Intense cold High Pressure systems over Canada and Eurasia create an induced Low over the Pole shown in Graphic 20. This effect is enhanced by the low level moisture and warmth that is absorbed by the atmosphere as cold air flows northward over (the comparatively) warmer Arctic Ocean water. These minus 40 to 70 degree Fahrenheit winds flow from the glaciers over upwelling warm 40 degrees Fahrenheit water creating an ideal situation for evaporating water from the ocean. The dry air vacuums moisture and warmth from the sea and the resulting air becomes much more unstable, this creates an ever thickening maritime layer.

In an Ice Age the open water surface of the Arctic Ocean create unstable air over them

Gale to storm force winds blow cold air northward from the intensely cold glaciers, as soon this air goes over upwelling water the extremely cold and dry air rapidly becomes modified and absorbs heat and moisture from the ocean surface. Storm force winds continually mix the surface layer of the Arctic Ocean, preventing the water from getting cold enough to form ice.

As storm force winds blow north from the glaciers, these winds drive now very saline and near freezing water towards the middle of the Arctic. These waters are now much cooler and as they move further north the winds increase and mixing increases in the surface layer. Finally in the center of the Arctic at the pole these now very saline and much colder waters combine and sink, becoming a strong underwater current that flows towards the Greenland / English Straits where the current flows into the Atlantic. To balance this outflow warm waters from the Atlantic flows northward in subsurface currents that bring warm Atlantic water into the Arctic Ocean. The cycle continues for 100,000 years.

Air moving northward into the Arctic is deflected to the right due to the spin of the world; Currents moving northward are likewise affected. A warm current moving northward from the Greenland Straits is likewise deflected toward the right; the warm water creating a subsurface current around the Arctic Ocean, resulting in a thick subsurface layer of warmer water moving counterclockwise around the periphery of the Arctic Ocean, this is upwelled due to the winds.

How does the Arctic become warm?

When the Arctic becomes clear of ice, the upwelling water around the periphery of the Arctic is warmer than the very cold glacier air masses that lay around it. This helps create an area of Low Pressure, as moisture and warmth from the open water creates unstable air that in turn creates a surface Low Pressure system. This sets the world into a cooling cycle; this lasted over 110,000 years in the last cycle. As soon as the Ice Age begins the world cools faster than at any other time in the cycle.

As the warm underlying water in the Arctic is drawn up by upwelling winds, the pool of warm underlying water becomes more and more shallow. It takes many thousands of years and eventually sea levels decrease several hundred feet and the inflow of warm water is not enough to balance the freight train like outflow of cold underwater currents carrying cold water out of the Arctic finally begins to slow.

A Storm is created

Two types of warm core Lows exist today, the thermal Low that forms over deserts of the world on summer days and tropical storms (such as Hurricanes or Typhoons). These two types of Low pressure systems are warmer towards the middle of the storm (hence the name).

In the atmosphere we have High pressure over warm surface Lows and vice a versa (this is nature's way of making things symmetrical, the exception being warm maritime highs). Warming in the lower atmosphere decreases surface pressures (creating cyclonic circulation which is counter clockwise in northern hemisphere) and increases pressures aloft (creating an upper level anti-cyclonic circulation). For instance a hurricane has cyclonic winds that bring moist air in at the surface but reverses wind direction near 24,000 feet and has a huge anticyclone outflow above, this is nature's way of maintaining balance. You can watch this on satellite loops of strong hurricane, a strong hurricane has lots of cirrus status outflow aloft above the lower level inflow, this is an indication its healthy and maintaining strength.

A Storm is created

1. The spin of the earth is most pronounced at the pole, this acts on any wind, causing it to turn to the right in its direction of movement.
2. Intense surface High pressure systems over Eurasia and Canada create an induced Low over the Arctic Ocean, basically an area of lower pressure between the two Highs.
3. Warm water flowing in from the Atlantic Ocean upwells to the surface around the Arctic Ocean and causes the air flowing over it to absorb moisture and heat, moist warm air is less dense than dry cold air and more unstable – it rises, helping to create a surface low.
4. The combination of the first three factors creates an intense Low pressure system centered on the North Pole.
5. Gale to storm force winds cause strong currents cause of intensely cold and salty water to come together and sink at the pole. This water eventually becomes an underwater current that flows towards straits and into the Atlantic.
6. This upwelling warm water and severe surface mixing of the ocean waters is why the Arctic does not freeze.

Bridges and Ice

Ice and cars do not mix
Roadway bridges get cold quicker than normal roadways
because they can cool from below as well as above
If it is near freezing or below freezing and liquid precipitation is falling
expect ice on bridges to happen
If you are driving and it starts icing up
Be smart
get a hotel and wait till it warms up or they clear the roads

CHAPTER 14

THE HURRICANE AND WHIRLPOOL AT THE TOP OF THE WORLD

My wildest idea (to date) and one that makes perfect sense is a hurricane forms at the pole, by far the largest and most organized storm on the planet. It is the result of a combination of an induced Low between two Highs and a warm core Low feeding off the moisture and heat from the Arctic Ocean's surface.

It is latent heat of condensation that fuels a hurricane, a complicated way of saying that when moisture condenses into water, heat is released; lots of heat. Think about how much energy is required to turn water into vapor (in a teapot over a burner); this same amount of heat is released when water vapor condenses back into water or snow. This is a secondary source of energy for hurricanes, it adds heat to the air that decreases density and causes more upward motion. No reason this same latent heat of condensation could not fuel a hurricane at the North Pole, it is just a lot cooler.

A warm Low pressure circulation over the Arctic is well organized, taking full effect of the world's spin. This helps create a nearly perfectly round warm Low pressure system at the pole that works in conjunction with storm force winds blowing ice cold air northward from the glaciers.

The surface outflow of dry and cold air (minus 40 to 80 degrees Fahrenheit) from the glaciers flows northward over the warm Arctic water and soaks up water like a sponge from the Arctic Ocean. This moisture laden air creates a deep maritime layer with multiple cloud layers and fog. This layer increases in depth the longer the air is over water, flowing northward and gradually turning to the right (due to the earth's rotation) and begins forming a massive vortex at the top of the world, centered on the North Pole.

The vortex at the top of the world looks like a hurricane with storm walls, precipitation, and high winds but remains stationary. It has nowhere to go, nor any reason to get there. Due to cooler temperatures in the overall air column the height of the storm is half the height of a hurricane in the tropics. The Arctic hurricane makes up for this lack of height by longevity, as it will be around for the best part of a 100,000 years.

Outside of the storm all surface winds flow northward turning towards the right (from the spin of the earth), flowing into the polar hurricane. In the lower 10,000 feet multiple layers of air enter the walls of the storm, some dry, some moist, others comparatively warm. It is the moist air that provides fuel to the storm, as it condenses into rain this releases a huge amount of energy called latent heat of condensation and is a primary source of energy for hurricanes.

Like a chimney for a low to be very efficient it needs outflow aloft, this vortex at the top of the world has an intense anticyclone aloft that carry's moisture away from the pole as a dense upper level outflow. This outflow is the source of moisture for glaciers building in the far north. These glaciers don't need much moisture to build, as they have plenty of time to grow.

This stratus and fog on the surface combined with dense upper level anti-cyclonic winds create insulating layers of clouds prevent heat from radiating out into space. This helps keep the Arctic warm.

What happens when you get storm force winds blowing in a circle over water for many thousands of years? The largest and most intense whirlpool ever in existence. How does this happen? Currents on the surface of the sea flow in the same direction as the dominate winds (subsurface currents get all screwy under it but that is another lesson).

Cold surface winds have evaporated huge amounts of water from the sea and these surface waters both increase in salinity and decrease in temperature. Normally these very saline and cold waters would sink, but in this situation the storm force winds creates monster seas and strong surface currents that carry these waters northward towards the pole where they converge in a huge whirlpool and sink.

Thus it is that in the center of the Hurricane at the top of the world we find a whirlpool larger and more intense than any ever seen; many miles across at the top. If anyone had traveled the shores of the Arctic Ocean during the last Ice Age it was highly dangerous, the winds and currents would carry them rapidly northward and they were sucked underwater. The waters within this whirlpool are very cold and salty and eventually turn into a strong underwater current flowing back towards the Atlantic sill in a dense subsurface current that carries the water out of the western portion of the Greenland Straits.

This is the cosmic sized crapper in the Arctic, debris from such a whirlpool that lasts for 100,000 years must cover the ocean bottom in layers thousands of feet deep.

Turbulence

Clear Air Turbulence

This is fast air moving over slower air and very dangerous

This often happens on the lee side of mountains
up to several hundred miles downwind,
kind of an eddy effect - with multiple eddies possible

Other Turbulence occurs around the Jet Stream
This is strongest on the east side of continents – when temperature differences are greatest
(that would be winter and spring)

ON ANY FLIGHT
ALWAYS WEAR A SEATBELT

CHAPTER 15

SUMMER OVER GLACIER REGIONS

When ice begins covering the northern continents, it affects the weather of the entire continent. It is easy to understand the winters and the continuous building of ice but why didn't the ice melt in the summer? When ice builds, one main affect is obvious in that ice and snow reflect over 80 % of incoming sunlight (even more at high latitudes), a huge amount compared to the 5 % of reflection of dark colored soil. When the sun does shine on the ice this reflectivity helps keep the ice from melting.

Air over glaciers is somewhat similar to the weather over San Diego, where a subsidence inversion prevails. The Pacific High is centered over a thousand miles west of California, on its right side cool surface air flows southward and has a lot of subsidence aloft (an effect on the right side of High pressure systems). This subsidence causes air to warm as it falls (adiabatic warming) and then it hits the top of the maritime layer. The surface maritime layer is cool and moist and the first thousand or two feet of air above the surface becomes distinctly different than the air above and dominated by stratus, stratocumulus, and fog; remaining cool in the day. An inversion at 1000-2000 feet protects the lower levels from the sun (this is the type of clouds that protected people during eruptions in the previously discussed Garden of Eden).

Air over glaciers have a very strong inversion and the same layers of ice fog and stratus prevents the surface layer from receiving direct sunlight. The surface layer over a glacier is very cold and as you go higher it gets warmer, until one reaches the top of the inversion layer and temperature starts becomes colder again.

In San Diego the stratus and fog of a maritime layer sometimes burn off in the heat of the day but only if enough heat is absorbed to break the inversion. For fog and stratus in San Diego this may be an increase of 15 or 20 degrees Fahrenheit, but for a glacier inversion the surface temperature of the ice might be a minus 40 to 80 degrees Fahrenheit and top of the inversion might be anywhere from 0 to 50 degrees Fahrenheit. For the sun to shine the sun must supply enough energy to break a 40 to 130 degree inversion, if this does not happen (and it rarely does), the clouds and fog continue to protect the ice from the sun.

Small areas of the glacier may clear (towards the southern edge), but much of the sun's heat during the day is reflected by ice and snow. Not enough heat is absorbed to melt the ice. Much ice melted along the southern regions of the glaciers during the short summers, but more ice is formed in the winters than melts in the summers.

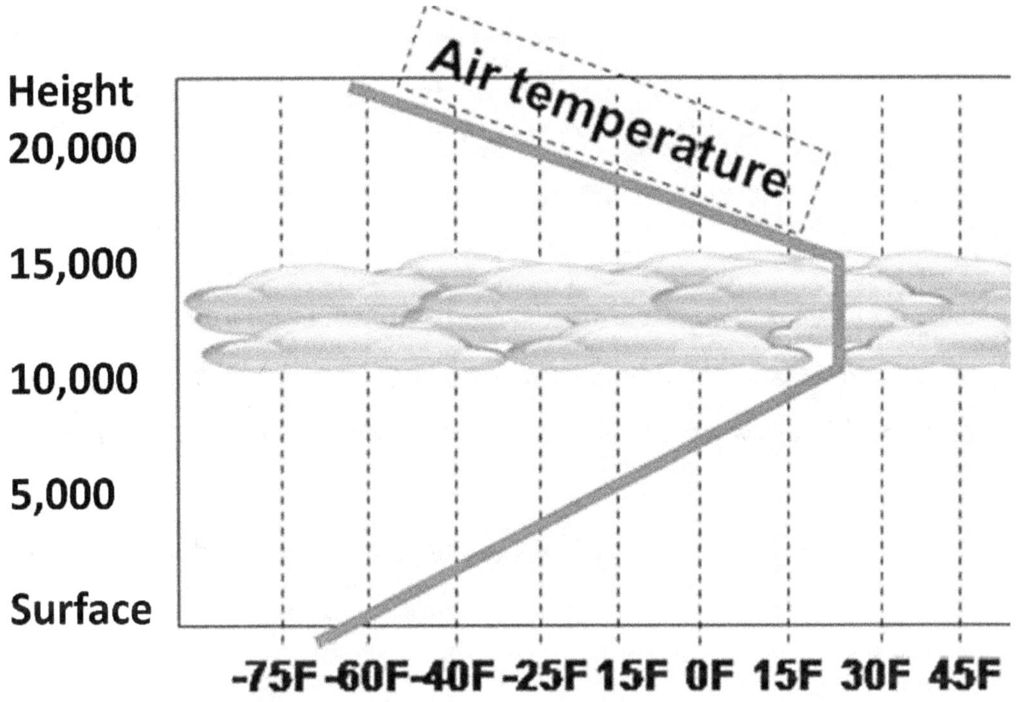

Graphic 21. The Inversion over a Glacier Air Mass

An Atmospheric Inversion

When very cold air has warmer air above it, it's called an inversion because the air gets warmer with height. This makes the surface very stable (and foggy in spots), it also creates a layer of status and stratocumulus at the height where the temperature begins falling again (typically 1,500 to 2,000 feet in height, an inversion over a glacier air mass is far higher).

An inversion is equivalent to having a heavy blanket protecting the ice, sunshine alone was not strong enough to break the icy grip of a glacier. Only the return of the Jet Stream (when the Arctic Ocean finally ices over again) which brings strong westerly winds, was warm enough to finally melt the cruel heart of the masses of ice.

Down Drafts

The air in a thunderstorm is moving violently
both
UPWARD and DOWNWARD

These areas of turbulence are dangerous anytime
As downdrafts of cold air can easily cause a jet liner to fly into the ground
(it happened in Dallas in 1985 when a landing plane encountered a microburst)

This is especially dangerous on takeoffs and landing
So when the plane is delayed due to thunderstorms

Relax and don't complain
They are just trying to keep you alive

CHAPTER 16

METHANE HYDRATE FORMATION

The Ice Record shows multiple instance of temperature rises (shown in Graphic 17), these short term temperature rises are caused by degassing of Methane Hydrate deposits on the continental shelves of the world's oceans. Methane Hydrate is conditionally stable above 500 pounds pressure and below 50 degrees Fahrenheit. Trillions of square feet of Methane Hydrate lay on the continental shelf's and slopes. Think of these deposits as trails of gunpowder as once any portion begins to degas the entire deposit may go off. Enough Methane Hydrate lays on the continental shelves' to significantly affect the world's climate for thousands of years (in the last ice age this repeatedly happened and each event caused about 3,000 years of warming).

In the ocean, water pressure rises 15 pounds per every 33 feet of depth. Hydrates at a depth of 1100 feet are under over 500 pounds of pressure. A loss of 400 feet of water above these Hydrates decreases the amount of water pressure on these Hydrates by 180 pounds. Hydrate under 320 pounds of pressure are a lot more unstable than before. Especially when accompanied with a significant increase in the sea water temperatures. This decrease in pressure and increase in temperature makes them liable to degas into bubbles and fresh water. Fresh water is less dense than salt water and degassing hydrates release lots of fresh water that rises to the surface carrying the gas bubbles of Methane with it; much of the gas is released into the atmosphere,

As large Methane Hydrate deposits degas, a large portion of the gas escapes into the atmosphere and gradually finds its way back into the ocean. A portion of the gas remains in the ocean and is captured again in a Hydrate matrix and re-deposited at lower depths. This cycle repeats over and over until sea levels have fallen as much as they can.

What effect does Methane have on the atmosphere? It is a much stronger greenhouse gas than Carbon Dioxide, over 33-72 (no one agrees on this) times more effective. Causing large increases in global temperatures, but this effect is mostly felt in areas away from the ice. The ice does feel some of the increase and a significant amount of the ice melts mostly near the southern edges, but not in the heart of the glaciers. Most of the heat accumulates in tropical and maritime air masses such as the Atlantic and Pacific Highs, ocean water temperatures also increase (some of this heat makes its way into the Arctic Ocean via currents). The overall affect is a stronger temperature gradient between the warm ocean air masses and the cold glacier air masses. This increases the amount and intensity of precipitation that falls over the ice and other areas, causing the glaciers to grow faster. These gradients are similar to today but much stronger.

During the depths of the Ice Age ocean levels varied by dozens of feet. This may have been due to variations in the sun's energy output, but the real culprit is likely the cyclic degassing of Methane Hydrate deposits on the continental shelves of the world's oceans. These degassing events raised worldwide temperatures and melted thousands of square miles of ice on the worlds land masses. This caused ocean levels to vary over thousands of years. Dozens of these warming events show up on the world's ice records.

Methane Hydrate formation

Visualize what happens in the deep ocean under the most nutrient rich areas, hundreds of feet of decaying material cover the ocean floor. From deep inside this mass of decaying material (where oxygen does not exist) Methane Gas forms and eventually enters the water.

Water compresses very little, while Methane gas compresses very well. These small Methane gas molecules create water masses of different density than surrounding water which changes the density of the water mass and this rises or lowers until it finds other water masses of equal density. This is where the Methane gas rich water builds up in layers, much like clouds in the atmosphere.

The deep ocean has weak currents that circulate bottom water and these carry these Methane gas rich water masses along and they eventually butt up against continental shelves where the current rises up the steep slopes and onto the continental shelves of the world's oceans. These slopes are often 15,000 feet deep and sometimes near vertical.

As currents flow up and over the lip of the continental shelf, the Methane gas rich water mass picks up particles of sand, clay, and fine calcite. These particles are a form of hydroscopic nuclei and where ice crystals start forming. It is the open spaces within developing ice crystal structures that Methane gas (trapped between water molecules) sees as a welcome escape from the surrounding intense water pressure and like college students on spring break, these gas molecules pack into these empty spaces and more crystal builds around them and lock them in. It is Hotel California all over again, they check in but can't check out; becoming locked into a crystal prison. This process continues and repeats until these crystals eventually grow so heavy that they fall to the ocean floor, looking very much like snow.

This reaction keeps happening because the currents keep a steady supply of Methane gas and hydroscopic particles moving into the formation area.

This trapped Methane gas is under the same pressure as the water around it, under 1,100 feet of water Methane is under 500 pounds of pressure and if the resulting hydrate were to degas at sea level, the gases within expand 33 times in size. If this hydrate was formed at 30,000 feet in depth these hydrates are under 13,000 pounds of pressure, this hydrate upon degassing at sea level pressure would expand over 910 times in size. This makes methane hydrate unattractive as an energy source, its unpredictable and expands either a little or a lot depending on the pressures under which it was created.

We do know that deposits of Methane Hydrate on the continental slopes are so large as to stagger the imagination, actually measured by square miles. Methane Hydrate looks like snow or ice and is literally a form of Methane locked in a complex ice matrix of water molecules. You can hold it in your hand and light it with a match, the bottom is cool while the top burns.

Perhaps someday we will mine it and use it to run our cars.

The Danger of Hydrates

These Methane Hydrates are conditionally stable, they are only stable under 500 pounds (or more) of pressure and in a cold environment (below 40 to 50 Fahrenheit). If either the pressure decreases or the temperature rises these Hydrates sometimes release the Methane they have held trapped for hundreds if not thousands of years.

Now consider that deposits of Hydrates on the sea floor are often a quarter mile deep. If such a deposit covered an area of 5 miles wide by 10 miles long (typical for small deposits on the continental shelf). When an area this size becomes unstable it expands at least 30 times. This results in a cloud of concentrated explosive gas over 30 miles in size. Methane likes to burn with a 14% concentration of methane to air this could expand and cover an area hundreds of miles in size, you just need some static and you get a nuclear fireball sized explosion.

The Tunguska event in Siberia in June 1908 was possibly caused by a mass release of Methane gas trapped under the thick frozen tundra that dominates the area. No meteor fragments or signature elements of a crater has ever been found in the area. They did find large some large deep holes that do not appear to be craters. These are similar to large holes with burnt edges hundreds of feet wide that have been appearing in Siberia over the last few years (these do not appear to be craters either). During the Tunguska event a bright fire trail was also seen in the sky and methane burns with a blue light.

The bogs in tundra are very deep and frozen many feet down, with layers of wet warm rotting vegetation in an oxygen free environment under hundreds of pounds of pressure far below the frozen layer (like a compost pile hundreds of miles wide). This lack of oxygen creates a perfect methane formation area and it becomes rich in Methane hydrates and the frozen tundra creates a perfect cap. The previous winter was very cold and the summer was warmer than normal. The thawing or an earthquake allowed the frozen tundra to crack and this released masses of the trapped high pressure Methane trapped under it. It was a clear dry day (perfect for generating static) and a spark could easily of caused an ignition source. Who knows? It's a theory of mine, I have a few.

We do live in a dangerous world, with many dangers that we never even considered. This is the increased Hydrate threat that man must face, as world temperatures increases. The threat of mass Methane Gas releases.

These mass releases of Hydrates under water also replace water with trillions of rising gas bubbles that reduce the ability of sea water to support a ships weight and may be the reason for the mysterious sinking of some ships.

It has been known for many years that ocean eddies form along the ocean currents, submarines have long used these eddies as hiding places as they trap sound waves and a submarine inside one is nearly invisible to hunting ships or planes. These eddies are often miles across and several thousand feet deep. Both warm and cold eddies form and affect large volumes of water. Every year hundreds of these eddies form along the major currents of the world and spin off to either side of the current. These eddies are strong and long lasting plus even at fifteen hundred feet in depth – temperatures might still be over 50 degrees Fahrenheit.

The threat to man is one of these huge warm eddies will drift over a major Methane Hydrate deposit. This could cause temperatures 1,000 feet down to increase enough that a spontaneous mass release of Methane occurs. As currents get much warmer, the great currents move more and warmer water and the eddies are deeper, warmer, and more frequent.

These massive Methane Hydrate deposits are connected and once a portion begins degassing, the entire deposit may degas like a trail of gunpowder going off. When a portion begins to degas, the reaction feeds on available hydrate deposits and goes on until all available hydrates have degassed. Imagine a gas cloud the size of New Jersey blowing over the coast, the two main dangers are suffocation and explosion. It could easily kill a small city, the ultimate gas cloud explosion.

If someone is trying to take your money

Be afraid

If they are doing something for YOUR own good

Be very afraid - RUN

CHAPTER 17

VOLCANIC ERUPTIONS

In 1991 I was a witness to the eruption of Pinatubo, a terrifying event where the eruptive column extended upwards over 20 miles, I know this because I was working in a weather office 21 miles away and we repeatedly tracked it on our radar, mesmerized by its strength and size. In one eruption over an 18 hour period, the volcano ejected 3 square kilometers of material. Small compared to the eruption of a super volcano such as Toba or Yellowstone that ejects a thousand times the material. These super volcanoes have huge underground reservoirs of magma under unimaginable pressure. Man has never seen one go off, we didn't even know they existed until the last hundred years or so.

Types of Caldera Volcano eruptions

- **Major:** These are blow outs, lakes of water above are instantly evaporated as the magma escapes the magma chamber below. The last big Yellowstone eruption was like this and occurred 600,000 years ago. Few of these events occur as minor eruptions are more frequent, these small ones are the way these massive killers release steam. A variation of this happened for Toba 75,000 years ago.

- **Minor:** Bernoulli's theorem states that if one constricts a pipe that pressure increases inside the pipe. This is a key in the second method of a caldera eruption. Imagine a bottle of shaken soda, if you simply pop the top off it flies and lands a few feet away and the soda escapes in a second or two, most landing 5 or 6 feet away (a major soda bottle eruption). Now poke a small hole in the top of the shaken soda, the soda is now going to spurt out in a long solid stream, for a minute or two. If the pin prick is in the top of a super volcano such as Toba or Yellowstone, the pressure in the vessel is in the millions of pounds and the volume is millions of square meters of magma. This creates a jet of magma than extends far into the stratosphere, a raging eruption that pumps ash and debris into the atmosphere. This starts a volcanic winter that could last for years.

Approximately 116,000 years ago, a volcanic event occurred that dropped global temperature in a way that can only be described as volcanic winter. This was a minor eruption of the Yellowstone Caldera (a pin prick) that lasted (on and off) for 8,000 years. Temperatures dropped faster than any time in the last 100,000 years, about 1 degree Celsius per 1,000 years. This drop in temperature occurred for 8,000 years straight and created hostile conditions for man that were not replicated until Toba Caldera erupted 30,000 years later (74,000 years ago).

The later eruption of Toba blamed for the near extermination of man did not produce nearly as abrupt or severe a drop in temperature. In fact the eruption of Toba, the second worst eruption in several million years is rather difficult to pick out on the ice record. It shows up but the temperature change is only slightly larger than later recurring Methane degassing events (that increase then decrease temperatures).

Fire & Ice

From the time our ancestors first swam out of the sea, the earth has tested the human race. Our survival tested by repeated trials of earthquake, flood, fire, and ice. During the bitter cold of the last Ice Age our race was very nearly exterminated by a combination of both fire and ice; 116,000 years ago Yellowstone erupted and around 74,000 years ago Toba Caldera Volcano erupted, both eruptions were capable of sending material almost into orbit. Toba threw ash 3,000 miles away in India. Winds don't normally carry ash so far and to cover an area this far away from the volcano with ash that was up to 40 feet deep shows how powerful Toba was. Magma was launched into the air with such force and violence as to be unimaginable. Maybe it was an extreme event that lasted for hours or a minor one that went on for weeks and years. We don't know.

Yellowstone is situated in such a way that an abundant amount of moisture is funneled into the area from the Pacific ocean. At the start of the last ice age this moisture created rapidly forming glaciers building right over the Yellowstone caldera. After 15,000 years the accumulated ice put much more stress on the magma below than it had been under prior to the Ice Age, resulting a series of minor eruptions that went on for several thousand years. Until the ice reached it maximum thickness, the pressure stopped rising, and the eruptions stopped.

Toba has chamber 20 miles down that holds a trillion square meters of magma and gases under at least 20 million pounds of pressure. Normal atmospheric pressure at the earth's surface is 15 pounds per square inch, during an eruption gases in a highly compressed magma chamber such as Toba literally expand hundreds of thousands of times as it reaches the surface. Such an eruption builds a magma column miles into the sky and nearly throw magma into orbit. Toba is very near the source of the 2004 Earthquake that generated the Christmas day Tsunami that killed 200,000 people. It is in all respects a man killer. We should watch this one closely.

The largest volcano in the solar system is on Mars and named Olympus Mons. Its base is the size of New Mexico and stands three times the height of Everest. It is pretty evident on satellite pictures. In the last few years quite a few meteorites have been found on Earth that came from Mars. Perhaps it was an eruption of Olympus Mons that launched these so violently they overcame the escape velocity of Mars (11,000 mph). This is conceivable as the atmosphere is pretty thin and friction a lot less that Earths.

Was the earthquake on Christmas 2004 a sign that Toba Caldera Volcano is waking up? It is impossible to tell, if it does mankind will again face possible extinction. But this time our population is in the billions and survival options are scarcer. The thought of an eruption of a caldera volcano capable of shooting a trillion square meters of magma into the atmosphere is terrifying, but not as terrifying as man's inability to deal with such an event. Look at what happened after Hurricane Katrina, our government is incapable of dealing with unforeseen events. I went through the aftermath of Pinatubo and it took months to get water and electricity back. Over a month to just clear the runways. An eruption of either Toba or Yellowstone could literally takes away the sun for months, disrupting the worlds protective layer of Ozone so when the sun does return it eventually blinds all who venture forth into the light of day (without eye protection) and kill most living plants.

When Pinatubo volcano erupted in 1991 about 5% of the ash that was ejected was magnetite, basically an iron ore that is magnetic. The ash clogged generator filters and jet engines, while the magnetite was attracted to electric motors, power lines, coils, open wires, etc… anything with a current or had iron content. It knocked down power grids which had to be completely cleaned before they were turned back on. Eruptions of either Toba or Yellowstone would affect power systems and electric motors within a 2,000 miles radius and 4,000 to 10,000 miles directly downwind (following prevailing upper level winds). No water, electricity, plants die; we get hungry and thirsty real fast.

In the long run civilization will not survive an eruption of a big caldera volcano, mankind would survive but civilization would crumble. Even in the short run it's a life or death situation for individuals, even my NRA buddy would run out of bullets. Only a few survive and all fall into savage survival mode, with survival going to the fiercest and those willing to eat whatever and whoever was at hand. Those who study DNA tell us that 74,000 years ago such a crisis occurred that only 1 in 10,000 fertile females survived to repopulate the species.

Civilization is a thin veneer, humanity has a great love for violence and blood, this was evident in Roman times as well as our own. Witness our crime rates, persistent love of violent sports, and how popular a theme death and destruction are in our movies and novels. These are clear signs of how little we have progressed from the days of barbarianism. Imagine mankind facing a catastrophic that literally takes away the sun for months, renders all communications except for landlines useless, grounds aircraft worldwide, interferes with GPS signals and all navigation is impossible, our electrical grids ash over and crash taking away most things we take for granted. We get thirsty and hungry fast. All exasperated by losing oil from the Middle East and a general end to ocean traffic. Ships can't navigate without GPS, they have to wait until the skies clear and who knows how long that takes. How far does the veneer of civilization carry in such an event? What about when the sun comes back and the protective Ozone layer is gone and ultraviolet radiation starts to blind people who survived and cause skin cancers and remaining plants start dying? Then winter sets in and it gets really cold all over the globe and stays that way for god knows how long. It certainly not going to be very long before the neighbors irritating dog becomes your dinner, this will be the beginning of the end, a battle to survive.

Volcano's come in four types

(Mikes Scale)

1. Scenic volcanoes that tourists come to watch (Kilauea).
2. Volcanoes that scare but don't do much damage or the damage is localized. These can destroy small towns: hundreds perish.
3. Eruptions that scare you so bad that you run away as fast as you can; Mt St Helens was about a 2.5-3, depending on where you were standing when she erupted. A 3 will kill for many miles around.
4. Eruptions that kill for hundreds of miles around and thousands of miles downwind, you are unable to run because you are already dead. These set off volcanic winters and are global killers; the effect of a class 4 eruption can kill around the globe.

A class 1 volcano has magma that has little to no compressed gas; this comes up to the surface with few surprises to anyone. Kilauea is such a volcano, erupting in slow motion, where a big eruption is measured in volume of magma that is released – not in how high the eruptive plume goes.

Toba and Yellowstone are class 4 volcanoes. You don't even want to be on the planet when these two volcanos decide to have a party and let go. They are the largest known volcanos in the world.
I suspect Nicaragua may harbor a larger one as Lake Nicaragua looks like a big caldera volcano. Both Toba and Yellowstone are caldera volcanoes with magma chambers several miles down that are terrifyingly huge, each holding over a trillion square meters of magma under millions of pounds of pressure.

Caldera volcanos do not have a cinder cone like a traditional volcanos such as Mount Fuji. No, these volcanoes blow up so explosively that they leave a huge depression in the earth, Toba's last eruption left a depression that is now a lake 60 miles long and 20 miles wide, with a growing island in the center of the lake that represents the pressure building up far below (Lake Nicaragua is bigger at 100 by 44 miles, with an island in the middle). Toba has a trillion meters of magma in its magma chamber. It seems as if these types of volcano have a different style of eruption – one with so much force and violence that most if not all of the ejected magma is launched into the stratosphere or beyond. Toba did leave lots of ash in the Indian Ocean and was 40 foot deep in portions of India.

During an Plinian eruption magma is forced up and out of the volcano under millions of pounds of pressure; this magma is ejected directly upwards at hundreds of miles an hour. The magma is forced upwards in a near solid flow and most of the highly compressed gases inside the magma column have not expanded much in size; this means that as material get higher in the column it just expands more. The magma near the outside of the magma column expands many times in size, as much as the gases are able to expand and quickly cool into a rigid shell. It builds its own shell as it reaches into the sky. The magma in the core of the magma column is still under millions of pounds of pressure and as the magma column rises upwards, the continuing expansion of gases inside the magma column forces it ever upwards, for miles.

In 1991, I had no idea what a Plinian eruption was and it would be years later before I researched it. It is astounding how high the magma column from a Plinian type eruption rises into the sky, I saw it with my own eye (could of done without it). Plinian eruptions are named after Pliny the Younger who recorded the events surrounding the eruption of Mount Vesuvius in AD 79, the death of his father in a rescue attempt during the eruption, and the destruction of Pompeii. These eruptions are terrifying. From our weather office in Subic Bay Philippines we watched Mount Pinatubo in 1991 go off the top of our radar scope on numerous occasions. It went straight up and did not flare out a bit.

Our weather radar was able to see up to 85,000 feet and the eruptive column was straight as a pencil all the way up, no flaring at all. We were 21 miles away and 20 minutes after every big eruption started we began to get sandy ash falling like rain along with some rocks up to half the size of your fist. Figure 25 miles up, 21 miles out, and 20 miles back down; that's 66 miles in 20 minutes. Magma in the column was rising straight up (to well over 100,000 feet) at over 350 miles an hour. Remember that as the ash fell, it was going terminal velocity of 100 miles an hour so it took 12 minutes just to fall. It was the ride up that was so fast. This was a weak volcano compared to Toba and Yellowstone,

Envision a water hose pointed upwards and it's under high pressure, the water only starts to spread out when it exits the hose. It no longer has the force of the water behind it pushing it forwards. Now think about water full of carbonated soda under high pressure, as it exits the hose it rises to a much higher height as the gases inside the flow expand.

Two things influence the building of a column of magma. One is the cooling on the outer sides of the column, these rapidly cool and become rigid, this acts like the rigid sides of the hose allowing the column to rise ever higher and higher into the sky. It creates its own structural support. This is the reason volcanic columns can reach many miles into the sky.

The other reason for the extreme height of Plinian eruptions is the magma is under huge amounts of pressure and contains large amounts of compressed gases. As the column rises its outer shell cools while the inner core of the column is still holding magma under very high pressure, gases in the magma in the core of the column continually expand and force the column ever higher. As the column of magma reaches higher and higher into the sky, the continuous expansion of gases inside the magma column continue to push the column ever upwards.

In an eruption, this pressure creates a rising tower of molten stone with a hard outer shell that could almost reach into space, it all depends on the amount of pressure the magma is originally under and this may be millions of pounds of pressure; every column of magma will reach its maximum height (entirely dependent on the amount of pressure it started with), which might be anywhere from 5 feet to 100 miles (we don't know how high). In such a magma column the weight of the column above is balanced by the pressure of the rising magma below and the internal pressure of the ever expanding magma inside the column.

A strong Plinian eruption creates a column of dense magma miles high, a column extending upward into the very stratosphere, a fountain of molten rock and gas that goes up as molten magma and comes down as ash – a substance that looks like and feels like sand. A column 1,000 feet across and 20 miles tall has many billions of square feet of magma suspended in the vertical column. When the eruption finally ceases, the suspended column falls back to the ground. This is a pyroclastic flow and there is a huge amount of material hitting the ground after it shoots its wad. You don't want to be near it when that happens.

An eruption of a caldera is similar to a huge rocket engine blasting material straight into the stratosphere. The eruption of Toba shot a plume several thousand miles northwest over India and to the northeast over the South China Sea off Vietnam. This eruption likely lasted weeks and sent hundreds of millions of square meters of magma into the Jet Streams of both the northern and southern hemispheres. Much of India was covered in Toda's ash, some areas very deep and other just a foot. India is over a million square miles, if only ½ of India got covered by a foot – that is over 20 trillion square feet of ash – thousands of miles away from the volcano. This was a major event.

A strong non-catastrophic eruption Yellowstone shoots ash and gases straight into the Jet Stream and quickly affects weather in the northern hemisphere. An eruption of Toba affects the entire world as it is right on the equator.

CHAPTER 18

CONCLUSION

We are close to the beginning of a new Ice Age Cycle, exactly how close is totally unknown. It will start fairly small but gradually gets bigger and affect more and more. Life will change for many people. It may take a twenty years of severe winters and a lot of snow and ice before the high pressure areas becomes permanently established, but once that is done we are in for a long hard winter. We have to be prepared in case it's the first year or two, as it's a complete unknown.

It's been a 130,000 years since the last start of an ice age and we don't have much more than Ice Core data to figure out how and why it started. Our scientist have nothing to grab hold of. It's been too long. Even the temperature data makes it difficult for them as the ice cores are more compressed the further down you do. Each razor scrape covers more and more years.

Abrupt weather changes have occurred four times in the past 400,000 years when the Earth was in just such a warm period as we are today. To me an ice age is imminent. Sadly it seems as if many eminent scholars are the last to admit this is a possibility at all. It is something that is just too difficult to prove, and they do have their reputations and all. It is definitely true that on a geological scale we are way past due for the start of a new Ice Age. Now it is only a matter of time: be it a 2021, ten years, or another century. The bottom line is that an ice free Arctic Ocean triggers the next slide into an Ice Age; I personally expect an Ice Age to begin the year after this event and after the last few years of strange weather I fear that it doesn't have to be fully free of ice for Tipping Point to occur, we don't know.
I don't know and I'm the only one that has broached this subject that I know of.

The ice record has shown us that transition into the next Ice Age comes very rapidly after the warm peak is reached. Granted this is rapid on a 100,000 year scale, not a 20 year scale. My contention is that once the Tipping point is reached we are on a downhill road. It is also true that one cold winter day in Canada and Eurasia brings a solid freeze and it will not thaw for 36 million days (100,000 years).

The first sign of the oncoming Ice Age (after the melting Arctic Ice pack) is an ever weakening Jet Stream, weakening and becoming more disorganized. We see that today.

Why is the Arctic ice melting? Where does the heat come from that is preventing the ice from forming as thick as it did the year before? The answer is simple, the heat energy is already in the Arctic Ocean and present under the cold surface water. This heat energy is coming from the Atlantic, flowing in as a warm current and becomes a warm layer of middle level water with colder water above it. Every year this layer of warm subsurface water is becoming larger and it just takes a melted ice pack and an early cold winter on the continents to start the next Ice Age.

The transition to an ice free Arctic Ocean may take a few more years, but when it is complete the Northern Hemisphere will drop into an Ice Age. The ice may take 20 years to get going but it isn't going anywhere after it falls. Over the last 400,000 years; transitions into four different Ice Ages have been very abrupt, coming right after the heating peak has been reached. It is very alarming because temperatures plummet for many thousands of years; it gets real cold, fast.

Unless you live in northeast or central Canada, Northeast US, or northern Europe (Eurasia), where the ice first begins building, I do not expect any change to be radical or life changing, at least not at first. Plus negative changes in one area result in positive changes elsewhere, the deserts in the American southwest, North Africa, and the Middle East will bloom into fertile land. Washington, Vancouver and Alaska will be pretty nice.

It is highly possible that we have already passed the Tipping Point as seen by our weakened Jet Stream. Only time will tell.

Disaster is just around the corner

Our children will see the start of an Ice Age, a time when ice comes to the north and does not melt, not in our lifetimes. It will take a to kick in but the eventual political and economic ramifications of such an event are enormous. Insurance companies go bankrupt from the loses and many families are left with what they can put in the truck, or what they can carry.

The very political stability of the world is undermined as populations of people from northern regions are forced to migrate southward by the ice and cold, southern regions may or may not greet these refugees from the north with open arms. Imagine the countries of old Russia deciding to head south into the Arab world (where much of the oil is) where they are certainly not welcome, nuclear weapons are used on every continent; war, starvation goes from being regional issues in third world countries, to being the norm - worldwide.

What regions are most at risk? That is the Great Lakes region and eastern Canada. I expect this area to freeze over in an ice storm and the ice stays. Summers in the region will be nonexistent for many years. Northern Europe and Russia are also at risk.

If it doesn't happen in 2021, it is still very close. Every sign I see points to it.

The notion that what is true today, may NOT be true tomorrow

Is a valid notion

. . . .

MJL

ABOUT THE AUTHOR

In January of 1999, I retired from the United State Navy as a Aerographers Mate (a weather forecaster), have since earned a Master's Degree in Business and a Bachelor Of Science Degree in Database Management. After retiring I started a second career in computers and for the last 16 years I have been a computer consultant to a variety of clients and continue doing that on a day to day basis. At least till recently, I am looking for a job.

I spent 20 years in the Navy, serving during the cold war of the 70,'s 80's and 90's. A curious career path for a liberal and pacifist, due to a combination of patriotism and desire to stay out of trouble, certainly was not for the money. I am a United States Navy Aerographer's Mate (retired now – but that is my rating) and proudly wore my uniform. I was a weather forecaster. Was a pleasure to do so and would do it again.

My career was spent on the other side of the Pacific, spending almost a quarter century in that special place where the vast Pacific Ocean and the sky come together. Specially trained to observe and understand how the ocean and atmosphere interact with each other; as the sun, ocean, and the atmosphere seek balance. Performing duties as a Weather Forecaster, Typhoon Forecaster, Anti-Submarine Warfare Analyst, and Navy Flight Weather Forecaster. Over 20 years plotting and analyzing surface, upper levels, and oceanographic data; and analyzing untold thousands of hourly satellite pictures. All the while intensely watching the Jet Stream over Asia, trying to decipher the deadly cold winter Siberian air mass over Mongolia and forecast its movement and effects on other air masses. Seeking clues on why it sometime sits for weeks in one place and other times begins to move. Learning firsthand of the bonds that tie these severely cold and dense air masses to the upper level Jet Stream and learning how they work as a team, one does not move without the other.

On why Ice Ages occur,

it is my thought that we don't have to look to the Milky Way for the answer.

Let us look at the ocean, air masses, and the Jet Stream for a cause.

In all your journeys

May you always have fair winds and following seas

Reference

Petit, J.R., et al., 2001 - Vostok Ice Core Data for 420,000 Years, IGBP PAGES/World Data Center for Paleoclimatology Data Contribution Series #2001-076. NOAA/NGDC Paleoclimatology Program, Boulder CO, USA.

www.ingramcontent.com/pod-product-compliance
Lightning Source LLC
Chambersburg PA
CBHW081352280526
45788CB00009B/2852
* 9 7 8 1 4 5 6 4 5 5 2 5 5 *